A First Course in Complex Analysis

Version 1.53

Matthias Beck
Department of Mathematics
San Francisco State University
San Francisco, CA 94132
mattbeck@sfsu.edu

Gerald Marchesi
Department of Mathematical Sciences
Binghamton University (SUNY)
Binghamton, NY 13902
marchesi@math.binghamton.edu

Dennis Pixton
Department of Mathematical Sciences
Binghamton University (SUNY)
Binghamton, NY 13902
dennis@math.binghamton.edu

Lucas Sabalka
Lincoln, NE 68502
sabalka@gmail.com

A First Course in Complex Analysis

Copyright 2002–2017 by the authors. All rights reserved. The most current version of this book is available at the website

http://math.sfsu.edu/beck/complex.html.

This book may be freely reproduced and distributed, provided that it is reproduced in its entirety from the most recent version. This book may not be altered in any way, except for changes in format required for printing or other distribution, without the permission of the authors.

This edition published with express permission of the authors by
Orthogonal Publishing L3C
Ann Arbor, Michigan
www.orthogonalpublishing.com

Typeset in Adobe Garamond Pro using Math Design mathematical fonts.

About the cover: The cover illustration, *Square Squared* by Robert Chaffer, shows two superimposed images. The foreground image represents the result of applying a transformation, $z \mapsto z^2$ (see Exercises 3.53 and 3.54), to the background image. The locally-conformable property of this mapping can be observed through matching the line segments, angles, and Sierpinski triangle features of the background image with their respective images in the foreground figure. (The foreground figure is scaled down to about 40% and repositioned to accommodate artistic and visibility considerations.)

The background image fills the square with vertices at 0, 1, $1 + i$, and i (the positive direction along the imaginary axis is chosen as downward.) It was prepared by using Michael Barnsley's chaos game, capitalizing on the fact that a square is self-tiling, and by using a fractal-coloring method. A subset of the image is seen as a standard Sierpinski triangle. The chaos game was also re-purposed to create the foreground image.

Robert Chaffer is Professor Emeritus at Central Michigan University. His academic interests are in abstract algebra, combinatorics, geometry, and computer applications. Since retirement from teaching he has devoted much of his time to applying those interests to creation of art images.
http://people.cst.cmich.edu/chaff1ra/Art_From_Mathematics/

"And what is the use of a book," thought Alice, "without pictures or conversations?"
Lewis Carroll (*Alice in Wonderland*)

About this book. *A First Course in Complex Analysis* was written for a one-semester undergraduate course developed at Binghamton University (SUNY) and San Francisco State University, and has been adopted at several other institutions. For many of our students, Complex Analysis is their first rigorous analysis (if not mathematics) class they take, and this book reflects this very much. We tried to rely on as few concepts from real analysis as possible. In particular, series and sequences are treated from scratch, which has the consequence that power series are introduced late in the course. The goal our book works toward is the Residue Theorem, including some nontraditional applications from both continuous and discrete mathematics.

A printed paperback version of this open textbook is available from Orthogonal Publishing (www.orthogonalpublishing.com) or your favorite online bookseller.

About the authors. Matthias Beck is a professor in the Mathematics Department at San Francisco State University. His research interests are in geometric combinatorics and analytic number theory. He is the author of two other books, *Computing the Continuous Discretely: Ingeger-point Enumeration in Polyhedra* (with Sinai Robins, Springer 2007) and *The Art of Proof: Basic Training for Deeper Mathematics* (with Ross Geoghegan, Springer 2010).

Gerald Marchesi is a lecturer in the Department of Mathematical Sciences at Binghamton University (SUNY).

Dennis Pixton is a professor emeritus in the Department of Mathematical Sciences at Binghamton University (SUNY). His research interests are in dynamical systems and formal languages.

Lucas Sabalka is an applied mathematician at a technology company in Lincoln, Nebraska. He works on 3-dimensional computer vision applications. He was formerly a professor of mathematics at St. Louis University, after postdoctoral positions at UC Davis and Binghamton University (SUNY). His mathematical research interests are in geometric group theory, low dimensional topology, and computational algebra.

A Note to Instructors. The material in this book should be more than enough for a typical semester-long undergraduate course in complex analysis; our experience taught us that there is more content in this book than fits into one semester. Depending on the nature of your course and its place in your department's overall curriculum, some sections can be either partially omitted or their definitions and theorems can be assumed true without delving into proofs. Chapter 10 contains optional longer homework problems that could also be used as group projects at the end of a course.

We would be happy to hear from anyone who has adopted our book for their course, as well as suggestions, corrections, or other comments.

Acknowledgements. We thank our students who made many suggestions for and found errors in the text. Special thanks go to Sheldon Axler, Collin Bleak, Pierre-Alexandre Bliman, Matthew Brin, John McCleary, Sharma Pallekonda, Joshua Palmatier, and Dmytro Savchuk for comments, suggestions, and additions after teaching from this book.

We thank Lon Mitchell for his initiative and support for the print version of our book with Orthogonal Publishing, and Bob Chaffer for allowing us to feature his art on the book's cover.

We are grateful to the American Institute of Mathematics for including our book in their Open Textbook Initiative (`aimath.org/textbooks`).

Contents

1	**Complex Numbers**	**1**
	1.1 Definitions and Algebraic Properties	2
	1.2 From Algebra to Geometry and Back	5
	1.3 Geometric Properties	9
	1.4 Elementary Topology of the Plane	12
2	**Differentiation**	**24**
	2.1 Limits and Continuity	24
	2.2 Differentiability and Holomorphicity	29
	2.3 The Cauchy–Riemann Equations	33
	2.4 Constant Functions	38
3	**Examples of Functions**	**44**
	3.1 Möbius Transformations	44
	3.2 Infinity and the Cross Ratio	47
	3.3 Stereographic Projection	51
	3.4 Exponential and Trigonometric Functions	56
	3.5 Logarithms and Complex Exponentials	59
4	**Integration**	**72**
	4.1 Definition and Basic Properties	72
	4.2 Antiderivatives	78
	4.3 Cauchy's Theorem	81
	4.4 Cauchy's Integral Formula	87
5	**Consequences of Cauchy's Theorem**	**98**
	5.1 Variations of a Theme	98
	5.2 Antiderivatives Again	101
	5.3 Taking Cauchy's Formulas to the Limit	103

6 Harmonic Functions — 111
6.1 Definition and Basic Properties 111
6.2 Mean-Value and Maximum/Minimum Principle 115

7 Power Series — 122
7.1 Sequences and Completeness 123
7.2 Series 126
7.3 Sequences and Series of Functions 132
7.4 Regions of Convergence 136

8 Taylor and Laurent Series — 147
8.1 Power Series and Holomorphic Functions 147
8.2 Classification of Zeros and the Identity Principle 153
8.3 Laurent Series 157

9 Isolated Singularities and the Residue Theorem — 170
9.1 Classification of Singularities 170
9.2 Residues 177
9.3 Argument Principle and Rouché's Theorem 181

10 Discrete Applications of the Residue Theorem — 189
10.1 Infinite Sums 189
10.2 Binomial Coefficients 190
10.3 Fibonacci Numbers 191
10.4 The Coin-Exchange Problem 192
10.5 Dedekind Sums 194

Appendix: Theorems from Calculus — 196

Solutions to Selected Exercises — 199

Index — 204

Chapter 1

Complex Numbers

Die ganzen Zahlen hat der liebe Gott geschaffen, alles andere ist Menschenwerk.
(God created the integers, everything else is made by humans.)
Leopold Kronecker (1823–1891)

The real numbers have many useful properties. There are operations such as addition, subtraction, and multiplication, as well as division by any nonzero number. There are useful laws that govern these operations, such as the commutative and distributive laws. We can take limits and do calculus, differentiating and integrating functions. But you cannot take a square root of -1; that is, you cannot find a real root of the equation

$$x^2 + 1 = 0. \qquad (1.1)$$

Most of you have heard that there is a "new" number i that is a root of (1.1); that is, $i^2 + 1 = 0$ or $i^2 = -1$. We will show that when the real numbers are enlarged to a new system called the *complex numbers*, which includes i, not only do we gain numbers with interesting properties, but we do not lose many of the nice properties that we had before.

The complex numbers, like the real numbers, will have the operations of addition, subtraction, multiplication, as well as division by any complex number except zero. These operations will follow all the laws that we are used to, such as the commutative and distributive laws. We will also be able to take limits and do calculus. And, there will be a root of (1.1).

As a brief historical aside, complex numbers did not originate with the search for a square root of -1; rather, they were introduced in the context of cubic equations. Scipione del Ferro (1465–1526) and Niccolò Tartaglia (1500–1557) discovered a way to find a root of any cubic polynomial, which was publicized by Gerolamo Cardano (1501–1576) and is often referred to as *Cardano's formula*. For the cubic polynomial $x^3 + px + q$, Cardano's formula involves the quantity $\sqrt{\frac{q^2}{4} + \frac{p^3}{27}}$. It is not hard to come up with examples for p and q for which the argument of this square root becomes negative and thus not computable within the real numbers. On the other hand (e.g., by arguing through the graph of a cubic polynomial), every cubic polynomial has at least one real root. This seeming contradiction can be solved using complex numbers, as was probably first exemplified by Rafael Bombelli (1526–1572).

In the next section we show exactly how the complex numbers are set up, and in the rest of this chapter we will explore the properties of the complex numbers. These properties will be of both algebraic (such as the commutative and distributive properties mentioned already) and geometric nature. You will see, for example, that multiplication can be described geometrically. In the rest of the book, the calculus of complex numbers will be built on the properties that we develop in this chapter.

1.1 Definitions and Algebraic Properties

There are many equivalent ways to think about a complex number, each of which is useful in its own right. In this section, we begin with a formal definition of a complex number. We then interpret this formal definition in more useful and easier-to-work-with algebraic language. Later we will see several more ways of thinking about complex numbers.

Definition. The *complex numbers* are pairs of real numbers,

$$\mathbb{C} := \{(x, y) : x, y \in \mathbb{R}\},$$

equipped with the *addition*

$$(x, y) + (a, b) := (x + a, y + b) \tag{1.2}$$

and the *multiplication*

$$(x, y) \cdot (a, b) := (xa - yb, xb + ya). \tag{1.3}$$

One reason to believe that the definitions of these binary operations are acceptable is that $\mathbb{C}$ is an extension of $\mathbb{R}$, in the sense that the complex numbers of the form $(x, 0)$ behave just like real numbers:

$$(x, 0) + (y, 0) = (x + y, 0) \quad \text{and} \quad (x, 0) \cdot (y, 0) = (xy, 0).$$

So we can think of the real numbers being embedded in $\mathbb{C}$ as those complex numbers whose second coordinate is zero.

The following result states the algebraic structure that we established with our definitions.

Proposition 1.1. $(\mathbb{C}, +, \cdot)$ is a field, that is, for all $(x, y), (a, b), (c, d) \in \mathbb{C}$:

$$(x, y) + (a, b) \in \mathbb{C} \tag{1.4}$$

$$\big((x, y) + (a, b)\big) + (c, d) = (x, y) + \big((a, b) + (c, d)\big) \tag{1.5}$$

$$(x, y) + (a, b) = (a, b) + (x, y) \tag{1.6}$$

$$(x, y) + (0, 0) = (x, y) \tag{1.7}$$

$$(x, y) + (-x, -y) = (0, 0) \tag{1.8}$$

$$(x, y) \cdot \big((a, b) + (c, d)\big) = (x, y) \cdot (a, b) + (x, y) \cdot (c, d) \tag{1.9}$$

$$(x, y) \cdot (a, b) \in \mathbb{C} \tag{1.10}$$

$$\big((x, y) \cdot (a, b)\big) \cdot (c, d) = (x, y) \cdot \big((a, b) \cdot (c, d)\big) \tag{1.11}$$

$$(x, y) \cdot (a, b) = (a, b) \cdot (x, y) \tag{1.12}$$

$$(x, y) \cdot (1, 0) = (x, y) \tag{1.13}$$

$$\text{for all } (x, y) \in \mathbb{C} \setminus \{(0, 0)\} : (x, y) \cdot \left(\tfrac{x}{x^2+y^2}, \tfrac{-y}{x^2+y^2}\right) = (1, 0) \tag{1.14}$$

What we are stating here can be compressed in the language of algebra: equations (1.4)–(1.8) say that $(\mathbb{C}, +)$ is an *Abelian group* with *unit element* $(0, 0)$; equations (1.10)–(1.14) say that $(\mathbb{C} \setminus \{(0, 0)\}, \cdot)$ is an Abelian group with unit element $(1, 0)$.

The proof of Proposition 1.1 is straightforward but nevertheless makes for good practice (Exercise 1.14). We give one sample:

Proof of (1.8). By our definition for complex addition and properties of additive inverses in $\mathbb{R}$,

$$(x, y) + (-x, -y) = (x + (-x), y + (-y)) = (0, 0). \qquad \square$$

The definition of our multiplication implies the innocent looking statement

$$(0, 1) \cdot (0, 1) = (-1, 0). \tag{1.15}$$

This identity together with the fact that

$$(a,0) \cdot (x,y) = (ax, ay)$$

allows an alternative notation for complex numbers. The latter implies that we can write

$$(x,y) = (x,0) + (0,y) = (x,0) \cdot (1,0) + (y,0) \cdot (0,1).$$

If we think—in the spirit of our remark about embedding $\mathbb{R}$ into $\mathbb{C}$—of $(x,0)$ and $(y,0)$ as the real numbers x and y, then this means that we can write any complex number (x,y) as a linear combination of $(1,0)$ and $(0,1)$, with the real coefficients x and y. Now $(1,0)$, in turn, can be thought of as the real number 1. So if we give $(0,1)$ a special name, say i, then the complex number that we used to call (x,y) can be written as $x \cdot 1 + y \cdot i$ or

$$x + iy.$$

Definition. The number x is called the *real part* and y the *imaginary part*[1] of the complex number $x + iy$, often denoted as $\operatorname{Re}(x + iy) = x$ and $\operatorname{Im}(x + iy) = y$.

The identity (1.15) then reads

$$i^2 = -1.$$

In fact, much more can now be said with the introduction of the square root of -1. It is not just that (1.1) has a root, but *every* nonconstant polynomial has roots in $\mathbb{C}$:

Fundamental Theorem of Algebra (see Theorem 5.11). *Every nonconstant polynomial of degree d has d roots (counting multiplicity) in $\mathbb{C}$.*

The proof of this theorem requires some (important) machinery, so we defer its proof and an extended discussion of it to Chapter 5.

We invite you to check that the definitions of our binary operations and Proposition 1.1 are coherent with the usual real arithmetic rules if we think of complex numbers as given in the form $x + iy$.

---

[1] The name has historical reasons: people thought of complex numbers as unreal, imagined.

1.2 From Algebra to Geometry and Back

Although we just introduced a new way of writing complex numbers, let's for a moment return to the (x, y)-notation. It suggests that we can think of a complex number as a two-dimensional real vector. When plotting these vectors in the plane $\mathbb{R}^2$, we will call the x-axis the *real axis* and the y-axis the *imaginary axis*. The addition that we defined for complex numbers resembles vector addition; see Figure 1.1. The analogy stops at multiplication: there is no "usual" multiplication of two vectors in $\mathbb{R}^2$ that gives another vector, and certainly not one that agrees with our definition of the product of two complex numbers.

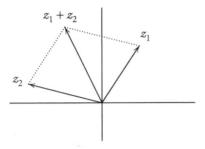

Figure 1.1: Addition of complex numbers.

Any vector in $\mathbb{R}^2$ is defined by its two coordinates. On the other hand, it is also determined by its length and the angle it encloses with, say, the positive real axis; let's define these concepts thoroughly.

Definition. The *absolute value* (also called the *modulus*) of $z = x + iy$ is

$$r = |z| := \sqrt{x^2 + y^2},$$

and an *argument* of $z = x + iy$ is a number $\varphi \in \mathbb{R}$ such that

$$x = r \cos\varphi \quad \text{and} \quad y = r \sin\varphi.$$

A given complex number $z = x + iy$ has infinitely many possible arguments. For instance, the number $1 = 1 + 0i$ lies on the positive real axis, and so has argument 0, but we could just as well say it has argument 2π, 4π, -2π, or $2\pi k$ for any integer k. The number $0 = 0 + 0i$ has modulus 0, and every real number φ is an argument.

Aside from the exceptional case of 0, for any complex number z, the arguments of z all differ by a multiple of 2π, just as we saw for the example $z = 1$.

The absolute value of the difference of two vectors has a nice geometric interpretation:

Proposition 1.2. Let $z_1, z_2 \in \mathbb{C}$ be two complex numbers, thought of as vectors in $\mathbb{R}^2$, and let $d(z_1, z_2)$ denote the *distance* between (the endpoints of) the two vectors in $\mathbb{R}^2$ (see Figure 1.2). Then

$$d(z_1, z_2) = |z_1 - z_2| = |z_2 - z_1|.$$

Proof. Let $z_1 = x_1 + iy_1$ and $z_2 = x_2 + iy_2$. From geometry we know that

$$d(z_1, z_2) = \sqrt{(x_1 - x_2)^2 + (y_1 - y_2)^2}.$$

This is the definition of $|z_1 - z_2|$. Since $(x_1 - x_2)^2 = (x_2 - x_1)^2$ and $(y_1 - y_2)^2 = (y_2 - y_1)^2$, this is also equal to $|z_2 - z_1|$. □

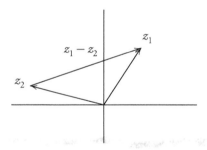

Figure 1.2: Geometry behind the distance between two complex numbers.

That $|z_1 - z_2| = |z_2 - z_1|$ simply says that the vector from z_1 to z_2 has the same length as the vector from z_2 to z_1.

It is very useful to keep this geometric interpretation in mind when thinking about the absolute value of the difference of two complex numbers.

One reason to introduce the absolute value and argument of a complex number is that they allow us to give a geometric interpretation for the multiplication of two complex numbers. Let's say we have two complex numbers: $x_1 + iy_1$, with absolute value r_1 and argument φ_1, and $x_2 + iy_2$, with absolute value r_2 and argument

φ_2. This means we can write $x_1 + iy_1 = (r_1 \cos\varphi_1) + i(r_1 \sin\varphi_1)$ and $x_2 + iy_2 = (r_2 \cos\varphi_2) + i(r_2 \sin\varphi_2)$. To compute the product, we make use of some classic trigonometric identities:

$$
\begin{aligned}
(x_1 + iy_1)(x_2 + iy_2) &= (r_1 \cos\varphi_1 + i\, r_1 \sin\varphi_1)(r_2 \cos\varphi_2 + i\, r_2 \sin\varphi_2) \\
&= (r_1 r_2 \cos\varphi_1 \cos\varphi_2 - r_1 r_2 \sin\varphi_1 \sin\varphi_2) + i(r_1 r_2 \cos\varphi_1 \sin\varphi_2 + r_1 r_2 \sin\varphi_1 \cos\varphi_2) \\
&= r_1 r_2 \big((\cos\varphi_1 \cos\varphi_2 - \sin\varphi_1 \sin\varphi_2) + i(\cos\varphi_1 \sin\varphi_2 + \sin\varphi_1 \cos\varphi_2)\big) \\
&= r_1 r_2 \big(\cos(\varphi_1 + \varphi_2) + i \sin(\varphi_1 + \varphi_2)\big).
\end{aligned}
$$

So the absolute value of the product is $r_1 r_2$ and one of its arguments is $\varphi_1 + \varphi_2$. Geometrically, we are multiplying the lengths of the two vectors representing our two complex numbers and adding their angles measured with respect to the positive real axis.[2]

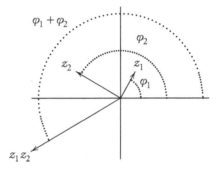

Figure 1.3: Multiplication of complex numbers.

In view of the above calculation, it should come as no surprise that we will have to deal with quantities of the form $\cos\varphi + i\sin\varphi$ (where φ is some real number) quite a bit. To save space, bytes, ink, etc., (and because "Mathematics is for lazy people"[3]) we introduce a shortcut notation and define

$$e^{i\varphi} := \cos\varphi + i\sin\varphi.$$

Figure 1.4 shows three examples. At this point, this exponential notation is indeed

---

[2] You should convince yourself that there is no problem with the fact that there are many possible arguments for complex numbers, as both cosine and sine are periodic functions with period 2π.

[3] Peter Hilton (Invited address, Hudson River Undergraduate Mathematics Conference 2000).

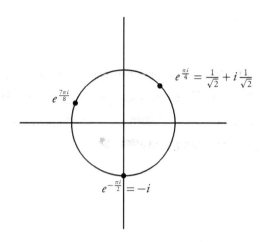

Figure 1.4: Three sample complex numbers of the form $e^{i\varphi}$.

purely a notation.[4] We will later see in Chapter 3 that it has an intimate connection to the complex exponential function. For now, we motivate this maybe strange seeming definition by collecting some of its properties:

Proposition 1.3. *For any $\varphi, \varphi_1, \varphi_2 \in \mathbb{R}$,*

(a) $e^{i\varphi_1} e^{i\varphi_2} = e^{i(\varphi_1 + \varphi_2)}$

(b) $e^{i0} = 1$

(c) $\frac{1}{e^{i\varphi}} = e^{-i\varphi}$

(d) $e^{i(\varphi + 2\pi)} = e^{i\varphi}$

(e) $|e^{i\varphi}| = 1$

(f) $\frac{d}{d\varphi} e^{i\varphi} = i e^{i\varphi}$.

You are encouraged to prove them (Exercise 1.16); again we give a sample.

Proof of ((f)). By definition of $e^{i\varphi}$,

$$\frac{d}{d\varphi} e^{i\varphi} = \frac{d}{d\varphi}(\cos\varphi + i\sin\varphi) = -\sin\varphi + i\cos\varphi = i(\cos\varphi + i\sin\varphi) = i e^{i\varphi}. \qquad \square$$

---

[4] In particular, while our notation "proves" *Euler's formula* $e^{2\pi i} = 1$, this simply follows from the facts $\sin(2\pi) = 0$ and $\cos(2\pi) = 1$. The connection between the numbers π, i, 1, and the complex exponential function (and thus the number e) is somewhat deeper. We'll explore this in Section 3.5.

Proposition 1.3 implies that $\left(e^{2\pi i \frac{m}{n}}\right)^n = 1$ for any integers m and $n > 0$. Thus numbers of the form $e^{2\pi i q}$ with $q \in \mathbb{Q}$ play a pivotal role in solving equations of the form $z^n = 1$—plenty of reason to give them a special name.

Definition. A *root of unity* is a number of the form $e^{2\pi i \frac{m}{n}}$ for some integers m and $n > 0$. Equivalently (by Exercise 1.17), a root of unity is a complex number ζ such that $\zeta^n = 1$ for some positive integer n. In this case, we call ζ an nth *root of unity*. If n is the smallest positive integer with the property $\zeta^n = 1$ then ζ is a *primitive* n^{th} root of unity.

Example 1.4. The 4th roots of unity are ± 1 and $\pm i = e^{\pm \frac{\pi i}{2}}$. The latter two are primitive 4th roots of unity. □

With our new notation, the sentence *the complex number $x + i y$ has absolute value r and argument φ* now becomes the identity

$$x + i y = r e^{i\varphi}.$$

The left-hand side is often called the *rectangular form*, the right-hand side the *polar form* of this complex number.

We now have five different ways of thinking about a complex number: the formal definition, in rectangular form, in polar form, and geometrically, using Cartesian coordinates or polar coordinates. Each of these five ways is useful in different situations, and translating between them is an essential ingredient in complex analysis. The five ways and their corresponding notation are listed in Figure 1.5. This list is not exhaustive; see, e.g., Exercise 1.21.

1.3 Geometric Properties

From the chain of basic inequalities $-\sqrt{x^2 + y^2} \leq -\sqrt{x^2} \leq x \leq \sqrt{x^2} \leq \sqrt{x^2 + y^2}$ (or, alternatively, by arguing with basic geometric properties of triangles), we obtain the inequalities

$$-|z| \leq \operatorname{Re}(z) \leq |z| \quad \text{and} \quad -|z| \leq \operatorname{Im}(z) \leq |z|. \tag{1.16}$$

The square of the absolute value has the nice property

$$|x + i y|^2 = x^2 + y^2 = (x + i y)(x - i y).$$

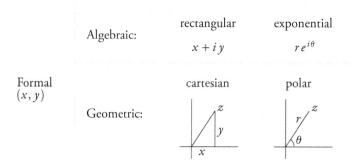

Figure 1.5: Five ways of thinking about a complex number.

This is one of many reasons to give the process of passing from $x + iy$ to $x - iy$ a special name.

Definition. The number $x - iy$ is the *(complex) conjugate* of $x + iy$. We denote the conjugate by
$$\overline{x + iy} := x - iy.$$

Geometrically, conjugating z means reflecting the vector corresponding to z with respect to the real axis. The following collects some basic properties of the conjugate.

Proposition 1.5. For any $z, z_1, z_2 \in \mathbb{C}$,

(a) $\overline{z_1 \pm z_2} = \overline{z_1} \pm \overline{z_2}$

(b) $\overline{z_1 \cdot z_2} = \overline{z_1} \cdot \overline{z_2}$

(c) $\overline{\left(\frac{z_1}{z_2}\right)} = \frac{\overline{z_1}}{\overline{z_2}}$

(d) $\overline{\overline{z}} = z$

(e) $|\overline{z}| = |z|$

(f) $|z|^2 = z\overline{z}$

(g) $\operatorname{Re}(z) = \frac{1}{2}(z + \overline{z})$

(h) $\operatorname{Im}(z) = \frac{1}{2i}(z - \overline{z})$

(i) $\overline{e^{i\varphi}} = e^{-i\varphi}$.

The proofs of these properties are easy (Exercise 1.22); once more we give a sample.

Proof of ((b)). Let $z_1 = x_1 + iy_1$ and $z_2 = x_2 + iy_2$. Then
$$\overline{z_1 \cdot z_2} = \overline{(x_1 x_2 - y_1 y_2) + i(x_1 y_2 + x_2 y_1)} = (x_1 x_2 - y_1 y_2) - i(x_1 y_2 + x_2 y_1)$$
$$= (x_1 - iy_1)(x_2 - iy_2) = \overline{z_1} \cdot \overline{z_2}. \qquad \square$$

We note that ((f)) yields a neat formula for the inverse of a nonzero complex number, which is implicit already in (1.14):

$$z^{-1} = \frac{1}{z} = \frac{\bar{z}}{|z|^2}.$$

A famous geometric inequality (which holds, more generally, for vectors in $\mathbb{R}^n$) goes as follows.

Proposition 1.6 (Triangle inequality). *For any $z_1, z_2 \in \mathbb{C}$ we have $|z_1 + z_2| \leq |z_1| + |z_2|$.*

By drawing a picture in the complex plane, you should be able to come up with a geometric proof of the triangle inequality. Here we proceed algebraically:

Proof. We make extensive use of Proposition 1.5:

$$\begin{aligned}|z_1 + z_2|^2 &= (z_1 + z_2)\overline{(z_1 + z_2)} = (z_1 + z_2)(\bar{z}_1 + \bar{z}_2) = z_1\bar{z}_1 + z_1\bar{z}_2 + z_2\bar{z}_1 + z_2\bar{z}_2 \\ &= |z_1|^2 + z_1\bar{z}_2 + \overline{z_1\bar{z}_2} + |z_2|^2 = |z_1|^2 + 2\operatorname{Re}(z_1\bar{z}_2) + |z_2|^2 \\ &\leq |z_1|^2 + 2|z_1\bar{z}_2| + |z_2|^2 = |z_1|^2 + 2|z_1||\bar{z}_2| + |z_2|^2 \\ &= |z_1|^2 + 2|z_1||z_2| + |z_2|^2 = (|z_1| + |z_2|)^2,\end{aligned}$$

where the inequality follows from (1.16). Taking square roots on the left- and right-hand side proves our claim. □

For future reference we list several useful variants of the triangle inequality:

Corollary 1.7. *For $z_1, z_2, \ldots, z_n \in \mathbb{C}$, we have the following relations:*

(a) *The triangle inequality:* $|\pm z_1 \pm z_2| \leq |z_1| + |z_2|$.

(b) *The reverse triangle inequality:* $|\pm z_1 \pm z_2| \geq ||z_1| - |z_2||$.

(c) *The triangle inequality for sums:*

$$\left|\sum_{k=1}^{n} z_k\right| \leq \sum_{k=1}^{n} |z_k|.$$

Inequality ((a)) is just a rewrite of the original triangle inequality, using the fact that $|\pm z| = |z|$, and ((c)) follows by induction. The proof of the reverse triangle inequality ((b)) is left as Exercise 1.25.

1.4 Elementary Topology of the Plane

In Section 1.2 we saw that the complex numbers $\mathbb{C}$, which were initially defined algebraically, can be identified with the points in the Euclidean plane $\mathbb{R}^2$. In this section we collect some definitions and results concerning the topology of the plane.

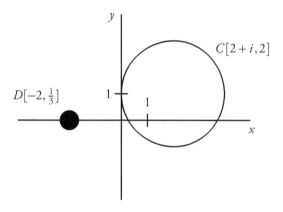

Figure 1.6: Sample circle and disk.

In Proposition 1.2, we interpreted $|z - w|$ as the distance between the complex numbers z and w, viewed as points in the plane. So if we fix a complex number a and a positive real number r, then all $z \in \mathbb{C}$ satisfying $|z - a| = r$ form the set of points at distance r from a; that is, this set is the *circle* with center a and radius r, which we denote by

$$C[a, r] := \{z \in \mathbb{C} : |z - a| = r\}.$$

The inside of this circle is called the *open disk* with center a and radius r; we use the notation

$$D[a, r] := \{z \in \mathbb{C} : |z - a| < r\}.$$

Note that $D[a, r]$ does not include the points on $C[a, r]$. Figure 1.6 illustrates these definitions.

Next we need some terminology for talking about subsets of $\mathbb{C}$.

Definition. Suppose G is a subset of $\mathbb{C}$.

(a) A point $a \in G$ is an *interior point* of G if some open disk with center a is a subset of G.

(b) A point $b \in \mathbb{C}$ is a *boundary point* of G if every open disk centered at b contains a point in G and also a point that is not in G.

(c) A point $c \in \mathbb{C}$ is an *accumulation point* of G if every open disk centered at c contains a point of G different from c.

(d) A point $d \in G$ is an *isolated point* of G if some open disk centered at d contains no point of G other than d.

The idea is that if you don't move too far from an interior point of G then you remain in G; but at a boundary point you can make an arbitrarily small move and get to a point inside G and you can also make an arbitrarily small move and get to a point outside G.

Definition. A set is *open* if all its points are interior points. A set is *closed* if it contains all its boundary points.

Example 1.8. For $r > 0$ and $a \in \mathbb{C}$, the sets $\{z \in \mathbb{C} : |z - a| < r\} = D[a, r]$ and $\{z \in \mathbb{C} : |z - a| > r\}$ are open. The *closed disk*

$$\overline{D}[a, r] := \{z \in \mathbb{C} : |z - a| \leq r\}$$

is an example of a closed set. $\square$

A given set might be neither open nor closed. The complex plane $\mathbb{C}$ and the *empty set* $\emptyset$ are (the only sets that are) both open and closed.

Definition. The *boundary* ∂G of a set G is the set of all boundary points of G. The *interior* of G is the set of all interior points of G. The *closure* of G is the set $G \cup \partial G$.

Example 1.9. The closure of the open disk $D[a, r]$ is $\overline{D}[a, r]$. The boundary of $D[a, r]$ is the circle $C[a, r]$. $\square$

Definition. The set G is *bounded* if $G \subseteq D[0, r]$ for some r.

One notion that is somewhat subtle in the complex domain is the idea of *connectedness*. Intuitively, a set is connected if it is "in one piece." In $\mathbb{R}$ a set is connected if and only if it is an interval, so there is little reason to discuss the matter. However, in the plane there is a vast variety of connected subsets.

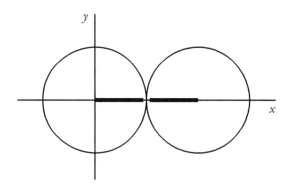

Figure 1.7: The intervals $[0, 1)$ and $(1, 2]$ are separated.

Definition. Two sets $X, Y \subseteq \mathbb{C}$ are *separated* if there are disjoint open sets $A, B \subset \mathbb{C}$ so that $X \subseteq A$ and $Y \subseteq B$. A set $G \subseteq \mathbb{C}$ is *connected* if it is impossible to find two separated nonempty sets whose union is G. A *region* is a connected open set.

The idea of separation is that the two open sets A and B ensure that X and Y cannot just "stick together." It is usually easy to check that a set is *not* connected. On the other hand, it is hard to use the above definition to show that a set *is* connected, since we have to rule out any possible separation.

Example 1.10. The intervals $X = [0, 1)$ and $Y = (1, 2]$ on the real axis are separated: There are infinitely many choices for A and B that work; one choice is $A = D[0, 1]$ and $B = D[2, 1]$, depicted in Figure 1.7. Hence $X \cup Y = [0, 2] \setminus \{1\}$ is not connected. □

One type of connected set that we will use frequently is a path.

Definition. A *path* (or *curve*) in $\mathbb{C}$ is a continuous function $\gamma : [a, b] \to \mathbb{C}$, where $[a, b]$ is a closed interval in $\mathbb{R}$. We may think of γ as a parametrization of the image that is painted by the path and will often write this parametrization as $\gamma(t)$, $a \leq t \leq b$.

The path is *smooth* if γ is differentiable and the derivative γ' is continuous and nonzero.[5]

This definition uses the calculus notions of continuity and differentiability; that is, $\gamma\colon [a,b] \to \mathbb{C}$ being *continuous* means that for all $t_0 \in [a,b]$

$$\lim_{t \to t_0} \gamma(t) = \gamma(t_0),$$

and the *derivative* of γ at t_0 is defined by

$$\gamma'(t_0) = \lim_{t \to t_0} \frac{\gamma(t) - \gamma(t_0)}{t - t_0}.$$

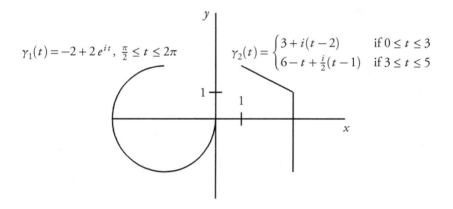

Figure 1.8: Two paths and their parametrizations; γ_1 is smooth and γ_2 is continuous and piecewise smooth.

Figure 1.8 shows two examples. We remark that each path comes with an *orientation*, i.e., a sense of direction. For example, the path γ_1 in Figure 1.8 is different from

$$\gamma_3(t) = -2 + 2e^{-it}, \quad 0 \leq t \leq \tfrac{3\pi}{2},$$

even though both γ_1 and γ_3 yield the same picture: γ_1 features a counter-clockwise orientation, where as that of γ_3 is clockwise.

---

[5] There is a subtlety here, because γ is defined on a closed interval. For $\gamma\colon [a,b] \to \mathbb{C}$ to be smooth, we demand both that $\gamma'(t)$ exists for all $a < t < b$, and that $\lim_{t \to a^+} \gamma'(t)$ and $\lim_{t \to b^-} \gamma'(t)$ exist.

It is a customary and practical abuse of notation to use the same letter for the path and its parametrization. We emphasize that a path must have a parametrization, and that the parametrization must be defined and continuous on a closed and bounded interval $[a, b]$. Since topologically we may identify $\mathbb{C}$ with $\mathbb{R}^2$, a path can be specified by giving two continuous real-valued functions of a real variable, $x(t)$ and $y(t)$, and setting $\gamma(t) = x(t) + i\, y(t)$.

Definition. The path $\gamma : [a, b] \to \mathbb{C}$ is *simple* if $\gamma(t)$ is one-to-one, with the possible exception that $\gamma(a) = \gamma(b)$ (in plain English: the path does not cross itself). A path $\gamma : [a, b] \to \mathbb{C}$ is *closed* if $\gamma(a) = \gamma(b)$.

Example 1.11. The *unit circle* $C[0, 1]$, parametrized, e.g., by $\gamma(t) = e^{it}$, $0 \le t \le 2\pi$, is a simple closed path. □

As seems intuitively clear, any path is connected; however, a proof of this fact requires a bit more preparation in topology. The same goes for the following result, which gives a useful property of open connected sets.

Theorem 1.12. If any two points in $G \subseteq \mathbb{C}$ can be connected by a path in G, then G is connected. Conversely, if $G \subseteq \mathbb{C}$ is open and connected, then any two points of G can be connected by a path in G; in fact, we can connect any two points of G by a chain of horizontal and vertical segments lying in G.

Here a *chain of segments* in G means the following: there are points $z_0, z_1, \ldots, z_n$ so that z_k and z_{k+1} are the endpoints of a horizontal or vertical segment in G, for all $k = 0, 1, \ldots, n-1$. (It is not hard to parametrize such a chain, so it determines a path.)

Example 1.13. Consider the open *unit disk* $D[0, 1]$. Any two points in $D[0, 1]$ can be connected by a chain of at most two segments in $D[0, 1]$, and so $D[0, 1]$ is connected. Now let $G = D[0, 1] \setminus \{0\}$; this is the punctured disk obtained by removing the center from $D[0, 1]$. Then G is open and it is connected, but now you may need more than two segments to connect points. For example, you need three segments to connect $-\frac{1}{2}$ to $\frac{1}{2}$ since we cannot go through 0. □

We remark that the second part of Theorem 1.12 is not generally true if G is not open. For example, circles are connected but there is no way to connect two distinct points of a circle by a chain of segments that are subsets of the circle. A more extreme example, discussed in topology texts, is the "topologist's sine curve," which is a connected set $S \subset \mathbb{C}$ that contains points that cannot be connected by a path of any sort within S.

Exercises

1.1. Let $z = 1 + 2i$ and $w = 2 - i$. Compute the following:

(a) $z + 3w$
(b) $\overline{w} - z$
(c) z^3
(d) $\text{Re}(w^2 + w)$
(e) $z^2 + \overline{z} + i$

1.2. Find the real and imaginary parts of each of the following:

(a) $\frac{z-a}{z+a}$ for any $a \in \mathbb{R}$
(b) $\frac{3+5i}{7i+1}$
(c) $\left(\frac{-1+i\sqrt{3}}{2}\right)^3$
(d) i^n for any $n \in \mathbb{Z}$

1.3. Find the absolute value and conjugate of each of the following:

(a) $-2 + i$
(b) $(2+i)(4+3i)$
(c) $\frac{3-i}{\sqrt{2}+3i}$
(d) $(1+i)^6$

1.4. Write in polar form:

(a) $2i$
(b) $1 + i$
(c) $-3 + \sqrt{3}i$
(d) $-i$
(e) $(2-i)^2$
(f) $|3 - 4i|$
(g) $\sqrt{5} - i$
(h) $\left(\frac{1-i}{\sqrt{3}}\right)^4$

1.5. Write in rectangular form:

(a) $\sqrt{2}e^{i\frac{3\pi}{4}}$
(b) $34 e^{i\frac{\pi}{2}}$
(c) $-e^{i250\pi}$
(d) $2e^{4\pi i}$

1.6. Write in both polar and rectangular form:

(a) $e^{\ln(5)i}$
(b) $\frac{d}{d\varphi} e^{\varphi + i\varphi}$

1.7. Show that the quadratic formula works. That is, for $a, b, c \in \mathbb{R}$ with $a \neq 0$, prove that the roots of the equation $az^2 + bz + c = 0$ are

$$\frac{-b \pm \sqrt{b^2 - 4ac}}{2a}.$$

Here we define $\sqrt{b^2 - 4ac} = i\sqrt{-b^2 + 4ac}$ if the *discriminant* $b^2 - 4ac$ is negative.

1.8. Use the quadratic formula to solve the following equations.

(a) $z^2 + 25 = 0$
(b) $2z^2 + 2z + 5 = 0$
(c) $5z^2 + 4z + 1 = 0$
(d) $z^2 - z = 1$
(e) $z^2 = 2z$

1.9. Find all solutions of the equation $z^2 + 2z + (1 - i) = 0$.

1.10. Fix $a \in \mathbb{C}$ and $b \in \mathbb{R}$. Show that the equation $|z^2| + \text{Re}(az) + b = 0$ has a solution if and only if $|a^2| \geq 4b$. When solutions exist, show the solution set is a circle.

1.11. Find all solutions to the following equations:

(a) $z^6 = 1$
(b) $z^4 = -16$
(c) $z^6 = -9$
(d) $z^6 - z^3 - 2 = 0$

1.12. Show that $|z| = 1$ if and only if $\frac{1}{z} = \overline{z}$.

1.13. Show that

(a) z is a real number if and only if $z = \overline{z}$;

(b) z is either real or purely imaginary if and only if $(\overline{z})^2 = z^2$.

1.14. Prove Proposition 1.1.

1.15. Show that if $z_1 z_2 = 0$ then $z_1 = 0$ or $z_2 = 0$.

1.16. Prove Proposition 1.3.

1.17. Fix a positive integer n. Prove that the solutions to the equation $z^n = 1$ are precisely $z = e^{2\pi i \frac{m}{n}}$ where $m \in \mathbb{Z}$. (*Hint*: To show that every solution of $z^n = 1$ is of this form, first prove that it must be of the form $z = e^{2\pi i \frac{a}{n}}$ for some $a \in \mathbb{R}$, then write $a = m + b$ for some integer m and some real number $0 \le b < 1$, and then argue that b has to be zero.)

1.18. Show that

$$z^5 - 1 = (z-1)\left(z^2 + 2z \cos \tfrac{\pi}{5} + 1\right)\left(z^2 - 2z \cos \tfrac{2\pi}{5} + 1\right)$$

and deduce from this closed formulas for $\cos \frac{\pi}{5}$ and $\cos \frac{2\pi}{5}$.

1.19. Fix a positive integer n and a complex number w. Find all solutions to $z^n = w$. (*Hint*: Write w in terms of polar coordinates.)

1.20. Use Proposition 1.3 to derive the triple angle formulas:

(a) $\cos(3\varphi) = \cos^3 \varphi - 3 \cos\varphi \sin^2 \varphi$

(b) $\sin(3\varphi) = 3 \cos^2 \varphi \sin\varphi - \sin^3 \varphi$

1.21. Given $x, y \in \mathbb{R}$, define the matrix $M(x, y) := \begin{bmatrix} x & y \\ -y & x \end{bmatrix}$. Show that

$$M(x, y) + M(a, b) = M(x + a, y + b)$$

and

$$M(x, y) M(a, b) = M(xa - yb, xb + ya).$$

(This means that the set $\{M(x, y) : x, y \in \mathbb{R}\}$, equipped with the usual addition and multiplication of matrices, behaves exactly like $\mathbb{C} = \{(x, y) : x, y \in \mathbb{R}\}$.)

1.22. Prove Proposition 1.5.

1.23. Sketch the following sets in the complex plane:

(a) $\{z \in \mathbb{C} : |z - 1 + i| = 2\}$

(b) $\{z \in \mathbb{C} : |z - 1 + i| \leq 2\}$

(c) $\{z \in \mathbb{C} : \operatorname{Re}(z + 2 - 2i) = 3\}$

(d) $\{z \in \mathbb{C} : |z - i| + |z + i| = 3\}$

(e) $\{z \in \mathbb{C} : |z| = |z + 1|\}$

(f) $\{z \in \mathbb{C} : |z - 1| = 2|z + 1|\}$

(g) $\{z \in \mathbb{C} : \operatorname{Re}(z^2) = 1\}$

(h) $\{z \in \mathbb{C} : \operatorname{Im}(z^2) = 1\}$

1.24. Suppose p is a polynomial with real coefficients. Prove that

(a) $\overline{p(z)} = p(\overline{z})$.

(b) $p(z) = 0$ if and only if $p(\overline{z}) = 0$.

1.25. Prove the reverse triangle inequality (Proposition 1.7((b))) $|z_1 - z_2| \geq |z_1| - |z_2|$.

1.26. Use the previous exercise to show that

$$\left| \frac{1}{z^2 - 1} \right| \leq \frac{1}{3}$$

for every z on the circle $C[0, 2]$.

1.27. Sketch the sets defined by the following constraints and determine whether they are open, closed, or neither; bounded; connected.

(a) $|z + 3| < 2$

(b) $|\operatorname{Im}(z)| < 1$

(c) $0 < |z - 1| < 2$

(d) $|z - 1| + |z + 1| = 2$

(e) $|z - 1| + |z + 1| < 3$

(f) $|z| \geq \operatorname{Re}(z) + 1$

1.28. What are the boundaries of the sets in the previous exercise?

1.29. Let G be the set of points $z \in \mathbb{C}$ satisfying either z is real and $-2 < z < -1$, or $|z| < 1$, or $z = 1$ or $z = 2$.

(a) Sketch the set G, being careful to indicate exactly the points that are in G.

(b) Determine the interior points of G.

(c) Determine the boundary points of G.

(d) Determine the isolated points of G.

1.30. The set G in the previous exercise can be written in three different ways as the union of two disjoint nonempty separated subsets. Describe them, and in each case say briefly why the subsets are separated.

1.31. Show that the union of two regions with nonempty intersection is itself a region.

1.32. Show that if $A \subseteq B$ and B is closed, then $\partial A \subseteq B$. Similarly, if $A \subseteq B$ and A is open, show that A is contained in the interior of B.

1.33. Find a parametrization for each of the following paths:

(a) the circle $C[1+i, 1]$, oriented counter-clockwise

(b) the line segment from $-1-i$ to $2i$

(c) the top half of the circle $C[0, 34]$, oriented clockwise

(d) the rectangle with vertices $\pm 1 \pm 2i$, oriented counter-clockwise

(e) the ellipse $\{z \in \mathbb{C} : |z-1| + |z+1| = 4\}$, oriented counter-clockwise

1.34. Let G be the annulus determined by the inequalities $2 < |z| < 3$. This is a connected open set. Find the maximum number of horizontal and vertical segments in G needed to connect two points of G.

Optional Lab

Open your favorite web browser and search for the *complex function grapher* for the open-source software *geogebra*.

1. Convert the following complex numbers into their polar representation, i.e., give the absolute value and the argument of the number.

$$34 = \qquad i = \qquad -\pi = \qquad 2 + 2i = \qquad -\frac{1}{2}\left(\sqrt{3} + i\right) =$$

   After you have finished computing these numbers, check your answers with the program.

2. Convert the following complex numbers given in polar representation into their rectangular representation.

$$2\,e^{i0} = \qquad 3\,e^{\frac{\pi i}{2}} = \qquad \frac{1}{2}e^{i\pi} = \qquad e^{-\frac{3\pi i}{2}} = \qquad 2\,e^{\frac{3\pi i}{2}} =$$

   After you have finished computing these numbers, check your answers with the program.

3. Pick your favorite five numbers from the ones that you've played around with and put them in the tables below, in both rectangular and polar form. Apply the functions listed to your numbers. Think about which representation is more helpful in each instance.

rectangular					
polar					
$z+1$					
$z+2-i$					
$2z$					
$-z$					
$\frac{z}{2}$					
iz					
$\overline{z}$					
z^2					
$\text{Re}(z)$					
$\text{Im}(z)$					
$i\,\text{Im}(z)$					
$|z|$					
$\frac{1}{z}$					

4. Play with other examples until you get a feel for these functions.

Chapter 2

Differentiation

> *Mathematical study and research are very suggestive of mountaineering. Whymper made several efforts before he climbed the Matterhorn in the 1860's and even then it cost the life of four of his party. Now, however, any tourist can be hauled up for a small cost, and perhaps does not appreciate the difficulty of the original ascent. So in mathematics, it may be found hard to realise the great initial difficulty of making a little step which now seems so natural and obvious, and it may not be surprising if such a step has been found and lost again.*
> Louis Joel Mordell (1888–1972)

We will now start our study of complex functions. *The* fundamental concept on which all of calculus is based is that of a limit—it allows us to develop the central properties of continuity and differentialbility of functions. Our goal in this chapter is to do the same for complex functions.

2.1 Limits and Continuity

Definition. A *(complex) function* f is a map from a subset $G \subseteq \mathbb{C}$ to $\mathbb{C}$; in this situation we will write $f : G \to \mathbb{C}$ and call G the *domain* of f. This means that each element $z \in G$ gets mapped to exactly one complex number, called the *image* of z and usually denoted by $f(z)$.

So far there is nothing that makes complex functions any more special than, say, functions from $\mathbb{R}^m$ to $\mathbb{R}^n$. In fact, we can construct many familiar looking functions from the standard calculus repertoire, such as $f(z) = z$ (the *identity map*), $f(z) = 2z + i$, $f(z) = z^3$, or $f(z) = \frac{1}{z}$. The former three could be defined on all of $\mathbb{C}$, whereas for the latter we have to exclude the origin $z = 0$ from the domain. On the other hand, we could construct some functions that make use of a certain representation of z, for example, $f(x, y) = x - 2iy$, $f(x, y) = y^2 - ix$, or $f(r, \varphi) = 2r \, e^{i(\varphi + \pi)}$.

Next we define limits of a function. The philosophy of the following definition is not restricted to complex functions, but for sake of simplicity we state it only for those functions.

Definition. Suppose $f : G \to \mathbb{C}$ and z_0 is an accumulation point of G. If w_0 is a complex number such that for every $\varepsilon > 0$ we can find $\delta > 0$ so that, for all $z \in G$ satisfying $0 < |z - z_0| < \delta$, we have $|f(z) - w_0| < \varepsilon$, then w_0 is the *limit* of f as z approaches z_0; in short,
$$\lim_{z \to z_0} f(z) = w_0.$$

This definition is the same as is found in most calculus texts. The reason we require that z_0 is an accumulation point of the domain is just that we need to be sure that there are points z of the domain that are arbitrarily close to z_0. Just as in the real case, our definition (i.e., the part that says $0 < |z - z_0|$) does not require that z_0 is in the domain of f and, if z_0 is in the domain of f, the definition explicitly ignores the value of $f(z_0)$.

Example 2.1. Let's prove that $\lim_{z \to i} z^2 = -1$.
Given $\varepsilon > 0$, we need to determine $\delta > 0$ such that $0 < |z - i| < \delta$ implies $|z^2 + 1| < \varepsilon$.
We rewrite
$$\left| z^2 + 1 \right| = |z - i||z + i| < \delta |z + i|.$$

If we choose δ, say, smaller than 1 then the factor $|z + i|$ on the right can be bounded by 3 (draw a picture!). This means that any $\delta < \min\{\frac{\varepsilon}{3}, 1\}$ should do the trick: in this case, $0 < |z - i| < \delta$ implies
$$\left| z^2 + 1 \right| < 3\delta < \varepsilon.$$

This was a proof written out in a way one might come up with it. Here's a short, neat version:

Given $\varepsilon > 0$, choose $0 < \delta < \min\{\frac{\varepsilon}{3}, 1\}$. Then $0 < |z - i| < \delta$ implies
$$|z + i| = |z - i + 2i| \leq |z - i| + |2i| < 3, \text{ so}$$
$$\left| z^2 - (-1) \right| = \left| z^2 + 1 \right| = |z - i||z + i| < 3\delta < \varepsilon.$$

This proves $\lim_{z \to i} z^2 = -1$. $\square$

Just as in the real case, the limit w_0 is unique if it exists (Exercise 2.3). It is often useful to investigate limits by restricting the way the point z approaches z_0. The following result is a direct consequence of the definition.

Proposition 2.2. Suppose $f : G \to \mathbb{C}$ and $\lim_{z \to z_0} f(z) = w_0$. Suppose $\widetilde{G} \subseteq G$ and z_0 is an accumulation point of $\widetilde{G}$. If $\widetilde{f}$ is the restriction of f to $\widetilde{G}$ then $\lim_{z \to z_0} \widetilde{f}(z)$ exists and has the value w_0.

The definition of *limit* in the complex domain has to be treated with a little more care than its real companion; this is illustrated by the following example.

Example 2.3. The limit of $\frac{\overline{z}}{z}$ as $z \to 0$ does not exist.

To see this, we try to compute this limit as $z \to 0$ on the real and on the imaginary axis. In the first case, we can write $z = x \in \mathbb{R}$, and hence

$$\lim_{z \to 0} \frac{\overline{z}}{z} = \lim_{x \to 0} \frac{\overline{x}}{x} = \lim_{x \to 0} \frac{x}{x} = 1.$$

In the second case, we write $z = iy$ where $y \in \mathbb{R}$, and then

$$\lim_{z \to 0} \frac{\overline{z}}{z} = \lim_{y \to 0} \frac{\overline{iy}}{iy} = \lim_{y \to 0} \frac{-iy}{iy} = -1.$$

So we get a different "limit" depending on the direction from which we approach 0. Proposition 2.2 then implies that the limit of $\frac{\overline{z}}{z}$ as $z \to 0$ does not exist. □

On the other hand, the following usual limit rules are valid for complex functions; the proofs of these rules are everything but trivial and make for nice practice (Exercise 2.4); as usual, we give a sample proof.

Proposition 2.4. Let f and g be complex functions with domain G, let z_0 be an accumulation point of G, and let $c \in \mathbb{C}$. If $\lim_{z \to z_0} f(z)$ and $\lim_{z \to z_0} g(z)$ exist, then

(a) $\lim_{z \to z_0} (f(z) + c\, g(z)) = \lim_{z \to z_0} f(z) + c \lim_{z \to z_0} g(z)$

(b) $\lim_{z \to z_0} (f(z) \cdot g(z)) = \lim_{z \to z_0} f(z) \cdot \lim_{z \to z_0} g(z)$

(c) $\lim_{z \to z_0} \frac{f(z)}{g(z)} = \frac{\lim_{z \to z_0} f(z)}{\lim_{z \to z_0} g(z)}$

where in the last identity we also require that $\lim_{z \to z_0} g(z) \neq 0$.

Proof of ((a)). Assume that $c \neq 0$ (otherwise there is nothing to prove), and let $L = \lim_{z \to z_0} f(z)$ and $M = \lim_{z \to z_0} g(z)$. Then we know that, given $\varepsilon > 0$, we can find $\delta_1, \delta_2 > 0$ such that

$$0 < |z - z_0| < \delta_1 \quad \text{implies} \quad |f(z) - L| < \frac{\varepsilon}{2}$$

and

$$0 < |z - z_0| < \delta_2 \quad \text{implies} \quad |g(z) - M| < \frac{\varepsilon}{2|c|}.$$

Thus, choosing $\delta = \min\{\delta_1, \delta_2\}$, we infer that $0 < |z - z_0| < \delta$ implies

$$|(f(z) + c\, g(z)) - (L + c\, M)| \leq |f(z) - L| + |c||g(z) - M| < \varepsilon.$$

Here we used the triangle inequality (Proposition 1.6). This proves $\lim_{z \to z_0}(f(z) + c\, g(z)) = L + c\, M$, which was our claim. □

Because the definition of the limit is somewhat elaborate, the following fundamental definition looks almost trivial.

Definition. Suppose $f : G \to \mathbb{C}$. If $z_0 \in G$ and either z_0 is an isolated point of G or

$$\lim_{z \to z_0} f(z) = f(z_0)$$

then f is *continuous* at z_0. More generally, f is *continuous* on $E \subseteq G$ if f is continuous at every $z \in E$.

However, in almost all proofs using continuity it is necessary to interpret this in terms of ε's and δ's. So here is an alternate definition:

Definition. Suppose $f : G \to \mathbb{C}$ and $z_0 \in G$. Then f is *continuous* at z_0 if, for every positive real number ε there is a positive real number δ so that

$$|f(z) - f(z_0)| < \varepsilon \quad \text{for all} \ z \in G \ \text{satisfying} \ |z - z_0| < \delta.$$

See Exercise 2.10 for a proof that these definitions are equivalent.

Example 2.5. We already proved (in Example 2.1) that the function $f : \mathbb{C} \to \mathbb{C}$ given by $f(z) = z^2$ is continuous at $z = i$. You're invited (Exercise 2.7) to extend our proof to show that, in fact, this function is continuous on $\mathbb{C}$.

On the other hand, let $g : \mathbb{C} \to \mathbb{C}$ be given by

$$g(z) := \begin{cases} \frac{\bar{z}}{z} & \text{if } z \neq 0, \\ 1 & \text{if } z = 0. \end{cases}$$

In Example 2.3 we proved that g is not continuous at $z = 0$. However, this is its only point of discontinuity (Exercise 2.8). □

Just as in the real case, we can "take the limit inside" a continuous function, by considering composition of functions.

Definition. The *image* of the function $g : G \to \mathbb{C}$ is the set $\{g(z) : z \in G\}$. If the image of g is contained in the domain of another function $f : H \to \mathbb{C}$, we define the *composition* $f \circ g : G \to \mathbb{C}$ through

$$(f \circ g)(z) := f(g(z)).$$

Proposition 2.6. Let $g : G \to \mathbb{C}$ with image contained in H, and let $f : H \to \mathbb{C}$. Suppose z_0 is an accumulation point of G, $\lim_{z \to z_0} g(z) = w_0 \in H$, and f is continuous at w_0. Then $\lim_{z \to z_0} f(g(z)) = f(w_0)$; in short,

$$\lim_{z \to z_0} f(g(z)) = f\left(\lim_{z \to z_0} g(z)\right).$$

Proof. Given $\varepsilon > 0$, we know there is an $\eta > 0$ such that

$$|w - w_0| < \eta \quad \text{implies} \quad |f(w) - f(w_0)| < \varepsilon.$$

For this η, we also know there is a $\delta > 0$ such that

$$0 < |z - z_0| < \delta \quad \text{implies} \quad |g(z) - w_0| < \eta.$$

Stringing these two implications together gives that

$$0 < |z - z_0| < \delta \quad \text{implies} \quad |f(g(z)) - f(w_0)| < \varepsilon.$$

We have thus proved that $\lim_{z \to z_0} f(g(z)) = f(w_0)$. □

2.2 Differentiability and Holomorphicity

The fact that simple functions such as $\frac{\bar{z}}{z}$ do not have limits at certain points illustrates something special about complex numbers that has no parallel in the reals—we can express a function in a very compact way in one variable, yet it shows some peculiar behavior in the limit. We will repeatedly notice this kind of behavior; one reason is that when trying to compute a limit of a function $f(z)$ as, say, $z \to 0$, we have to allow z to approach the point 0 in any way. On the real line there are only two directions to approach 0—from the left or from the right (or some combination of those two). In the complex plane, we have an additional dimension to play with. This means that the statement *A complex function has a limit ...* is in many senses stronger than the statement *A real function has a limit ...* This difference becomes apparent most baldly when studying derivatives.

Definition. Suppose $f : G \to \mathbb{C}$ is a complex function and z_0 is an interior point of G. The *derivative of f at z_0* is defined as

$$f'(z_0) := \lim_{z \to z_0} \frac{f(z) - f(z_0)}{z - z_0}, \tag{2.1}$$

provided this limit exists. In this case, f is called *differentiable* at z_0. If f is differentiable for all points in an open disk centered at z_0 then f is called *holomorphic*[1] at z_0. The function f is *holomorphic* on the open set $E \subseteq G$ if it is differentiable (and hence holomorphic) at every point in E. Functions that are differentiable (and hence holomorphic) in the whole complex plane $\mathbb{C}$ are called *entire*.

Example 2.7. The function $f : \mathbb{C} \to \mathbb{C}$ given by $f(z) = z^3$ is entire, that is, holomorphic in $\mathbb{C}$: For any $z_0 \in \mathbb{C}$,

$$\lim_{z \to z_0} \frac{f(z) - f(z_0)}{z - z_0} = \lim_{z \to z_0} \frac{z^3 - z_0^3}{z - z_0} = \lim_{z \to z_0} \frac{(z^2 + z z_0 + z_0^2)(z - z_0)}{z - z_0} = 3 z_0^2 .$$

$\square$

---

[1] Some sources use the term *analytic* instead of *holomorphic*. As we will see in Chapter 8, in our context, these two terms are synonymous. Technically, though, these two terms have different definitions. Since we will be using the above definition, we will stick with using the term *holomorphic* instead of the term *analytic*.

The difference quotient limit (2.1) can be rewritten as

$$f'(z_0) = \lim_{h \to 0} \frac{f(z_0 + h) - f(z_0)}{h}.$$

This equivalent definition is sometimes easier to handle. Note that h need not be a real number but can rather approach zero from anywhere in the complex plane.

The notions of differentiability and holomorphicity are not interchangeable:

Example 2.8. The function $f : \mathbb{C} \to \mathbb{C}$ given by $f(z) = (\bar{z})^2$ is differentiable at 0 and nowhere else; in particular, f is *not* holomorphic at 0: Let's write $z = z_0 + r\, e^{i\varphi}$. Then

$$\frac{\bar{z}^2 - \bar{z}_0^2}{z - z_0} = \frac{\overline{(z_0 + r\, e^{i\varphi})}^2 - \bar{z}_0^2}{z_0 + r\, e^{i\varphi} - z_0} = \frac{(\bar{z}_0 + r\, e^{-i\varphi})^2 - \bar{z}_0^2}{r\, e^{i\varphi}}$$

$$= \frac{\bar{z}_0^2 + 2\bar{z}_0\, r\, e^{-i\varphi} + r^2 e^{-2i\varphi} - \bar{z}_0^2}{r\, e^{i\varphi}} = \frac{2\bar{z}_0\, r\, e^{-i\varphi} + r^2 e^{-2i\varphi}}{r\, e^{i\varphi}}$$

$$= 2\bar{z}_0\, e^{-2i\varphi} + r\, e^{-3i\varphi}.$$

If $z_0 \neq 0$ then taking the limit of $f(z)$ as $z \to z_0$ thus means taking the limit of $2\bar{z}_0\, e^{-2i\varphi} + r\, e^{-3i\varphi}$ as $r \to 0$, which gives $2\bar{z}_0\, e^{-2i\varphi}$, a number that depends on φ, i.e., on the direction that z approaches z_0. Hence this limit does not exist.

On the other hand, if $z_0 = 0$ then the right-hand side above equals $r\, e^{-3i\varphi} = |z|\, e^{-3i\varphi}$. Hence

$$\lim_{z \to 0} \left| \frac{\bar{z}^2}{z} \right| = \lim_{z \to 0} \left| |z|\, e^{-3i\varphi} \right| = \lim_{z \to 0} |z| = 0,$$

which implies, by Exercise 2.5, that

$$\lim_{z \to z_0} \frac{\bar{z}^2 - \bar{z}_0^2}{z - z_0} = \lim_{z \to 0} \frac{\bar{z}^2}{z} = 0. \qquad \square$$

Example 2.9. The function $f : \mathbb{C} \to \mathbb{C}$ given by $f(z) = \bar{z}$ is nowhere differentiable:

$$\lim_{z \to z_0} \frac{\bar{z} - \bar{z}_0}{z - z_0} = \lim_{z \to z_0} \frac{\overline{z - z_0}}{z - z_0} = \lim_{z \to 0} \frac{\bar{z}}{z},$$

which does not exist, as discussed in Example 2.3. $\qquad \square$

The basic properties for derivatives are similar to those we know from real calculus. In fact, the following rules follow mostly from properties of the limit.

Proposition 2.10. Suppose f and g are differentiable at $z \in \mathbb{C}$ and h is differentiable at $g(z)$. Then

(a) $\bigl(f(z) + c\, g(z)\bigr)' = f'(z) + c\, g'(z)$ for any $c \in \mathbb{C}$

(b) $\bigl(f(z) g(z)\bigr)' = f'(z) g(z) + f(z) g'(z)$

(c) $\left(\dfrac{f(z)}{g(z)}\right)' = \dfrac{f'(z) g(z) - f(z) g'(z)}{g(z)^2}$ provided that $g(z)^2 \neq 0$

(d) $(z^n)' = n z^{n-1}$ for any nonzero integer n

(e) g is continuous at z

(f) $\bigl(h(g(z))\bigr)' = h'(g(z)) g'(z)$.

Proof of ((b)).

$$\begin{aligned}
\bigl(f(z)g(z)\bigr)' &= \lim_{h \to 0} \frac{f(z+h) g(z+h) - f(z) g(z)}{h} \\
&= \lim_{h \to 0} \frac{f(z+h)(g(z+h) - g(z)) + (f(z+h) - f(z)) g(z)}{h} \\
&= \lim_{h \to 0} f(z+h) \frac{g(z+h) - g(z)}{h} + \lim_{h \to 0} \frac{f(z+h) - f(z)}{h} g(z) \\
&= f(z) g'(z) + f'(z) g(z).
\end{aligned}$$

Note that we have used the definition of the derivative and Proposition 2.4((a)) & ((b)) (the addition and multiplication rules for limits). □

A prominent application of the differentiation rules is the composition of a complex function $f(z)$ with a path $\gamma(t)$. The proof of the following result gives a preview.

Proposition 2.11. Suppose f is holomorphic at $a \in \mathbb{C}$ with $f'(a) \neq 0$ and suppose γ_1 and γ_2 are two smooth paths that pass through a, making an angle of φ with each other. Then f transforms γ_1 and γ_2 into smooth paths which meet at $f(a)$, and the transformed paths make an angle of φ with each other.

In words, a holomorphic function with nonzero derivative preserves angles. Functions that preserve angles in this way are called *conformal*.

Proof. Let $\gamma_1(t)$ and $\gamma_2(t)$ be parametrizations of the two paths such that $\gamma_1(0) = \gamma_2(0) = a$. Then $\gamma_1'(0)$ (considered as a vector) is the tangent vector of γ_1 at the point a, and $\gamma_2'(0)$ is the tangent vector of γ_2 at a. Moving to the image of γ_1 and γ_2 under f, the tangent vector of $f(\gamma_1)$ at the point $f(a)$ is

$$\frac{d}{dt}f(\gamma_1(t))\Big|_{t=0} = f'(\gamma_1(0))\gamma_1'(0) = f'(a)\gamma_1'(0),$$

and similarly, the tangent vector of $f(\gamma_2)$ at the point $f(a)$ is $f'(a)\gamma_2'(0)$. This means that the action of f multiplies the two tangent vectors $\gamma_1'(0)$ and $\gamma_2'(0)$ by the same nonzero complex number $f'(a)$, and so the two tangent vectors got dilated by $|f'(a)|$ (which does not affect their direction) and rotated by the same angle (an argument of $f'(a)$). □

We end this section with yet another differentiation rule, that for inverse functions. As in the real case, this rule is only defined for functions that are bijections.

Definition. A function $f : G \to H$ is *one-to-one* if for every image $w \in H$ there is a unique $z \in G$ such that $f(z) = w$. The function is *onto* if every $w \in H$ has a preimage $z \in G$ (that is, there exists $z \in G$ such that $f(z) = w$). A *bijection* is a function that is both one-to-one and onto. If $f : G \to H$ is a bijection then $g : H \to G$ is the *inverse of* f if $f(g(z)) = z$ for all $z \in H$; in other words, the composition $f \circ g$ is the identity function on H.

Proposition 2.12. Suppose $G, H \subseteq \mathbb{C}$ are open sets, $f : G \to H$ is a bijection, $g : H \to G$ is the inverse function of f, and $z_0 \in H$. If f is differentiable at $g(z_0)$ with $f'(g(z_0)) \neq 0$ and g is continuous at z_0, then g is differentiable at z_0 with

$$g'(z_0) = \frac{1}{f'(g(z_0))}.$$

Proof. Since $f(g(z)) = z$ for all $z \in H$,

$$g'(z_0) = \lim_{z \to z_0} \frac{g(z) - g(z_0)}{z - z_0}$$
$$= \lim_{z \to z_0} \frac{g(z) - g(z_0)}{f(g(z)) - f(g(z_0))} = \lim_{z \to z_0} \frac{1}{\frac{f(g(z)) - f(g(z_0))}{g(z) - g(z_0)}}.$$

Now define $w_0 = g(z_0)$ and set

$$\varphi(w) := \begin{cases} \frac{f(w) - f(w_0)}{w - w_0} & \text{if } w \neq w_0 \\ f'(w_0) & \text{if } w = w_0. \end{cases}$$

This is continuous at w_0 and $\lim_{z \to z_0} g(z) = w_0$ by continuity of g, so we can apply Proposition 2.6:

$$g'(z_0) = \lim_{z \to z_0} \frac{1}{\varphi(g(z))} = \frac{1}{\varphi\left(\lim_{z \to z_0} g(z)\right)} = \frac{1}{f'(w_0)} = \frac{1}{f'(g(z_0))}. \qquad \square$$

2.3 The Cauchy–Riemann Equations

When considering a real-valued function $f : \mathbb{R}^2 \to \mathbb{R}$ of two variables, there is no notion of *the* derivative of a function. For such a function, we instead only have partial derivatives $\frac{\partial f}{\partial x}(x_0, y_0)$ and $\frac{\partial f}{\partial y}(x_0, y_0)$ (and also directional derivatives) which depend on the way in which we approach a point $(x_0, y_0) \in \mathbb{R}^2$. For a complex-valued function $f(z)$, we now have a new concept of the derivative $f'(z_0)$, which by definition cannot depend on the way in which we approach a point $z_0 = (x_0, y_0) \in \mathbb{C}$. It is logical, then, that there should be a relationship between the complex derivative $f'(z_0)$ and the *partial derivatives*

$$\frac{\partial f}{\partial x}(z_0) := \lim_{x \to x_0} \frac{f(x, y_0) - f(x_0, y_0)}{x - x_0}$$

and

$$\frac{\partial f}{\partial y}(z_0) := \lim_{y \to y_0} \frac{f(x_0, y) - f(x_0, y_0)}{y - y_0}$$

(so this definition is exactly as in the real-valued case). This relationship between the complex derivative and partial derivatives is very strong, and it is a powerful computational tool. It is described by the *Cauchy–Riemann equations*, named after Augustin Louis Cauchy (1789–1857) and Georg Friedrich Bernhard Riemann (1826–1866), even though the equations appeared already in the works of Jean le Rond d'Alembert (1717–1783) and Leonhard Euler (1707–1783).

Theorem 2.13. (a) Suppose f is differentiable at $z_0 = x_0 + i y_0$. Then the partial derivatives of f exist and satisfy

$$\frac{\partial f}{\partial x}(z_0) = -i \frac{\partial f}{\partial y}(z_0). \tag{2.2}$$

(b) Suppose f is a complex function such that the partial derivatives $\frac{\partial f}{\partial x}$ and $\frac{\partial f}{\partial y}$ exist in an open disk centered at z_0 and are continuous at z_0. If these partial derivatives satisfy (2.2) then f is differentiable at z_0.

In both cases (a) and (b), f' is given by

$$f'(z_0) = \frac{\partial f}{\partial x}(z_0).$$

Before proving Theorem 2.13, we note several comments and give two applications. It is traditional, and often convenient, to write the function f in terms of its real and imaginary parts. That is, we write $f(z) = f(x, y) = u(x, y) + i v(x, y)$ where u is the real part of f and v is the imaginary part. Then, using the usual shorthand $f_x = \frac{\partial f}{\partial x}$ and $f_y = \frac{\partial f}{\partial y}$,

$$f_x = u_x + i v_x \qquad \text{and} \qquad -i f_y = -i(u_y + i v_y) = v_y - i u_y.$$

With this terminology we can rewrite (2.2) as the pair of equations

$$\begin{aligned} u_x(x_0, y_0) &= v_y(x_0, y_0) \\ u_y(x_0, y_0) &= -v_x(x_0, y_0). \end{aligned} \tag{2.3}$$

As stated, parts (a) and (b) in Theorem 2.13 are not quite converse statements. However, we will show in Corollary 5.5 that if f is *holomorphic* at $z_0 = x_0 + i y_0$ then u and v have continuous partials (of any order) at z_0. That is, in Chapter 5 we will see that $f = u + iv$ is holomorphic in an open set G if and only if u and v have continuous partials that satisfy (2.3) in G.

If u and v satisfy (2.3) and their second partials are also continuous, then

$$u_{xx}(x_0, y_0) = v_{yx}(x_0, y_0) = v_{xy}(x_0, y_0) = -u_{yy}(x_0, y_0), \qquad (2.4)$$

that is,

$$u_{xx}(x_0, y_0) + u_{yy}(x_0, y_0) = 0$$

(and an analogous identity for v). Functions with continuous second partials satisfying this partial differential equation on a region $G \subset \mathbb{C}$ (though not necessarily (2.3)) are called *harmonic* on G; we will study such functions in Chapter 6. Again, as we will see later, if f is holomorphic in an open set G then the partials of any order of u and v exist; hence we will show that the real and imaginary parts of a function that is holomorphic in an open set are harmonic on that set.

Example 2.14. We revisit Example 2.7 and again consider $f : \mathbb{C} \to \mathbb{C}$ given by

$$f(z) = z^3 = (x + iy)^3 = \left(x^3 - 3xy^2\right) + i\left(3x^2 y - y^3\right).$$

Thus

$$f_x(z) = 3x^2 - 3y^2 + 6ixy \qquad \text{and} \qquad f_y(z) = -6xy + 3ix^2 - 3iy^2$$

are continuous on $\mathbb{C}$ and satisfy $f_x = -i f_y$. Thus by Theorem 2.13(b), $f(z) = z^3$ is entire. $\square$

Example 2.15. Revisiting Example 2.8 (you saw that coming, didn't you?), we consider $f : \mathbb{C} \to \mathbb{C}$ given by

$$f(z) = (\overline{z})^2 = (x - iy)^2 = x^2 - y^2 - 2ixy.$$

Now

$$f_x(z) = 2x - 2iy \qquad \text{and} \qquad f_y(z) = -2y - 2ix,$$

which satisfy $f_x = -i f_y$ only when $z = 0$. (The contrapositive of) Theorem 2.13(a) thus implies that $f(z) = (\overline{z})^2$ is not differentiable on $\mathbb{C} \setminus \{0\}$. $\square$

Proof of Theorem 2.13. (a) If f is differentiable at $z_0 = (x_0, y_0)$ then

$$f'(z_0) = \lim_{\Delta z \to 0} \frac{f(z_0 + \Delta z) - f(z_0)}{\Delta z}.$$

As we know by now, we must get the same result if we restrict Δz to be on the real axis and if we restrict it to be on the imaginary axis. In the first case, $\Delta z = \Delta x$ and

$$f'(z_0) = \lim_{\Delta x \to 0} \frac{f(z_0 + \Delta x) - f(z_0)}{\Delta x}$$
$$= \lim_{\Delta x \to 0} \frac{f(x_0 + \Delta x, y_0) - f(x_0, y_0)}{\Delta x} = \frac{\partial f}{\partial x}(x_0, y_0).$$

In the second case, $\Delta z = i \Delta y$ and

$$f'(z_0) = \lim_{i \Delta y \to 0} \frac{f(z_0 + i \Delta y) - f(z_0)}{i \Delta y}$$
$$= \lim_{\Delta y \to 0} \frac{1}{i} \frac{f(x_0, y_0 + \Delta y) - f(x_0, y_0)}{\Delta y} = -i \frac{\partial f}{\partial y}(x_0, y_0).$$

Thus we have shown that $f'(z_0) = f_x(z_0) = -i f_y(z_0)$.

(b) Suppose the Cauchy–Riemann equation (2.2) holds and the partial derivatives f_x and f_y are continuous in an open disk centered at z_0. Our goal is to prove that $f'(z_0) = f_x(z_0)$. By (2.2),

$$f_x(z_0) = \frac{\Delta x + i \Delta y}{\Delta z} f_x(z_0) = \frac{\Delta x}{\Delta z} f_x(z_0) + \frac{\Delta y}{\Delta z} i f_x(z_0) = \frac{\Delta x}{\Delta z} f_x(z_0) + \frac{\Delta y}{\Delta z} f_y(z_0).$$

On the other hand, we can rewrite the difference quotient for $f'(z_0)$ as

$$\frac{f(z_0 + \Delta z) - f(z_0)}{\Delta z} = \frac{f(z_0 + \Delta z) - f(z_0 + \Delta x) + f(z_0 + \Delta x) - f(z_0)}{\Delta z}$$
$$= \frac{f(z_0 + \Delta x + i \Delta y) - f(z_0 + \Delta x)}{\Delta z} + \frac{f(z_0 + \Delta x) - f(z_0)}{\Delta z}.$$

Thus

$$\lim_{\Delta z \to 0} \frac{f(z_0 + \Delta z) - f(z_0)}{\Delta z} - f_x(z_0)$$
$$= \lim_{\Delta z \to 0} \frac{\Delta y}{\Delta z} \left(\frac{f(z_0 + \Delta x + i\Delta y) - f(z_0 + \Delta x)}{\Delta y} - f_y(z_0) \right)$$
$$+ \lim_{\Delta z \to 0} \frac{\Delta x}{\Delta z} \left(\frac{f(z_0 + \Delta x) - f(z_0)}{\Delta x} - f_x(z_0) \right). \quad (2.5)$$

We claim that both limits on the right-hand side are 0, so we have achieved our set goal. The fractions $\frac{\Delta x}{\Delta z}$ and $\frac{\Delta y}{\Delta z}$ are bounded in absolute value by 1, so we just need to see that the limits of the expressions in parentheses are 0. The second term on the right-hand side of (2.5) has a limit of 0 since, by definition,

$$f_x(z_0) = \lim_{\Delta x \to 0} \frac{f(z_0 + \Delta x) - f(z_0)}{\Delta x}$$

and taking the limit here as $\Delta z \to 0$ is the same as taking the limit as $\Delta x \to 0$.

We cannot do something equivalent for the first term in (2.5), since now both Δx and Δy are involved, and both change as $\Delta z \to 0$. Instead we apply the Mean-Value Theorem A.2 for real functions,[2] to the real and imaginary parts $u(z)$ and $v(z)$ of $f(z)$. Theorem A.2 gives real numbers $0 < a, b < 1$ such that

$$\frac{u(x_0 + \Delta x, y_0 + \Delta y) - u(x_0 + \Delta x, y_0)}{\Delta y} = u_y(x_0 + \Delta x, y_0 + a\Delta y)$$
$$\frac{v(x_0 + \Delta x, y_0 + \Delta y) - v(x_0 + \Delta x, y_0)}{\Delta y} = v_y(x_0 + \Delta x, y_0 + b\Delta y).$$

---

[2] We collect several theorems from calculus, such as the Mean-Value Theorem for real-valued functions, in the Appendix.

Thus

$$\frac{f(z_0 + \Delta x + i\,\Delta y) - f(z_0 + \Delta x)}{\Delta y} - f_y(z_0)$$

$$= \left(\frac{u(x_0 + \Delta x, y_0 + \Delta y) - u(x_0 + \Delta x, y_0)}{\Delta y} - u_y(z_0)\right)$$

$$+ i\left(\frac{v(x_0 + \Delta x, y_0 + \Delta y) - v(x_0 + \Delta x, y_0)}{\Delta y} - v_y(z_0)\right)$$

$$= \left(u_y(x_0 + \Delta x, y_0 + a\,\Delta y) - u_y(x_0, y_0)\right) + i\left(v_y(x_0 + \Delta x, y_0 + b\,\Delta y) - v_y(x_0, y_0)\right). \tag{2.6}$$

Because u_y and v_y are continuous at (x_0, y_0),

$$\lim_{\Delta z \to 0} u_y(x_0 + \Delta x, y_0 + a\,\Delta y) = u_y(x_0, y_0)$$

and

$$\lim_{\Delta z \to 0} v_y(x_0 + \Delta x, y_0 + b\,\Delta y) = v_y(x_0, y_0),$$

and so (2.6) goes to 0 as $\Delta z \to 0$, which we set out to prove. $\square$

2.4 Constant Functions

As a sample application of the definition of the derivative of a complex function, we consider functions that have a derivative of 0. In a typical calculus course, one of the first applications of the Mean-Value Theorem for real-valued functions (Theorem A.2) is to show that if a function has zero derivative everywhere on an interval then it must be constant.

Proposition 2.16. If I is an interval and $f : I \to \mathbb{R}$ is a real-valued function with $f'(x)$ defined and equal to 0 for all $x \in I$, then there is a constant $c \in \mathbb{R}$ such that $f(x) = c$ for all $x \in I$.

Proof. The Mean-Value Theorem A.2 says that for any $x, y \in I$,

$$f(y) - f(x) = f'(x + a(y - x))(y - x)$$

for some $0 < a < 1$. Now $f'(x + a(y - x)) = 0$, so the above equation yields $f(y) = f(x)$. Since this is true for any $x, y \in I$, the function f must be constant on I. □

We do not (yet) have a complex version of the Mean-Value Theorem, and so we will use a different argument to prove that a complex function whose derivative is always 0 must be constant.

Our proof of Proposition 2.16 required two key features of the function f, both of which are somewhat obviously necessary. The first is that f be differentiable everywhere in its domain. In fact, if f is not differentiable everywhere, we can construct functions that have zero derivative almost everywhere but that have infinitely many values in their image.

The second key feature is that the interval I is connected. It is certainly important for the domain to be connected in both the real and complex cases. For instance, if we define the function $f : \{x + iy \in \mathbb{C} : x \neq 0\} \to \mathbb{C}$ through

$$f(z) := \begin{cases} 1 & \text{if } \operatorname{Re} z > 0, \\ 2 & \text{if } \operatorname{Re} z < 0, \end{cases}$$

then $f'(z) = 0$ for all z in the domain of f, but f is not constant. This may seem like a silly example, but it illustrates a pitfall to proving a function is constant that we must be careful of. Recall that a *region* of $\mathbb{C}$ is an open connected subset.

Theorem 2.17. *If $G \subseteq \mathbb{C}$ is a region and $f : G \to \mathbb{C}$ is a complex-valued function with $f'(z)$ defined and equal to 0 for all $z \in G$, then f is constant.*

Proof. We will first show that f is constant along horizontal segments and along vertical segments in G.

Suppose that H is a horizontal line segment in G. Thus there is some number $y_0 \in \mathbb{R}$ such that the imaginary part of any $z \in H$ is $\operatorname{Im}(z) = y_0$. Now consider the real part $u(z)$ of the function $f(z)$, for $z \in H$. Since $\operatorname{Im}(z) = y_0$ is constant on H, we can consider $u(z) = u(x, y_0)$ to be just a function of x, the real part of $z = x + iy_0$. By assumption, $f'(z) = 0$, so for $z \in H$ we have $u_x(z) = \operatorname{Re}(f'(z)) = 0$. Thus, by Proposition 2.16, $u(z)$ is constant on H.

We can argue the same way to see that the imaginary part $v(z)$ of $f(z)$ is constant on H, since $v_x(z) = \operatorname{Im}(f'(z)) = 0$ on H. Since both the real and imaginary parts of $f(z)$ are constant on H, the function $f(z)$ itself is constant on H.

This same argument works for vertical segments, interchanging the roles of the real and imaginary parts. We have thus proved that f is constant along horizontal segments and along vertical segments in G. Now if x and y are two points in G that can be connected by a path composed of horizontal and vertical segments, we conclude that $f(x) = f(y)$. But any two points of a region may be connected by finitely many such segments by Theorem 1.12, so f has the same value at any two points of G, thus proving the theorem. □

There are a number of surprising applications of Theorem 2.17; see, e.g., Exercises 2.19 and 2.20.

Exercises

2.1. Use the definition of limit to show for any $z_0 \in \mathbb{C}$ that $\lim_{z \to z_0} (az+b) = az_0 + b$.

2.2. Evaluate the following limits or explain why they don't exist.

(a) $\lim_{z \to i} \frac{iz^3 - 1}{z + i}$

(b) $\lim_{z \to 1-i} (x + i(2x + y))$

2.3. Prove that, if a limit exists, then it is unique.

2.4. Prove Proposition 2.4.

2.5. Let $f : G \to \mathbb{C}$ and suppose z_0 is an accumulation point of G. Show that $\lim_{z \to z_0} f(z) = 0$ if and only if $\lim_{z \to z_0} |f(z)| = 0$.

2.6. Proposition 2.2 is useful for showing that limits **do not** exist, but it is not at all useful for showing that a limit **does** exist. For example, define

$$f(z) = \frac{x^2 y}{x^4 + y^2} \quad \text{where} \quad z = x + iy \neq 0.$$

Show that the limits of f at 0 along all straight lines through the origin exist and are equal, but $\lim_{z \to 0} f(z)$ does not exist. (*Hint*: Consider the limit along the parabola $y = x^2$.)

2.7. Show that the function $f : \mathbb{C} \to \mathbb{C}$ given by $f(z) = z^2$ is continuous on $\mathbb{C}$.

2.8. Show that the function $g : \mathbb{C} \to \mathbb{C}$ given by

$$g(z) = \begin{cases} \frac{\bar{z}}{z} & \text{if } z \neq 0, \\ 1 & \text{if } z = 0 \end{cases}$$

is continuous on $\mathbb{C} \setminus \{0\}$.

2.9. Determine where each of the following functions $f : \mathbb{C} \to \mathbb{C}$ is continuous:

(a) $f(z) = \begin{cases} 0 & \text{if } z = 0 \text{ or } |z| \text{ is irrational,} \\ \frac{1}{q} & \text{if } |z| = \frac{p}{q} \in \mathbb{Q} \setminus \{0\} \text{ (written in lowest terms).} \end{cases}$

(b) $f(z) = \begin{cases} 0 & \text{if } z = 0, \\ \sin\varphi & \text{if } z = r e^{i\varphi} \neq 0. \end{cases}$

2.10. Show that the two definitions of continuity in Section 2.1 are equivalent. Consider separately the cases where z_0 is an accumulation point of G and where z_0 is an isolated point of G.

2.11. Consider the function $f : \mathbb{C} \setminus \{0\} \to \mathbb{C}$ given by $f(z) = \frac{1}{z}$. Apply the definition of the derivative to give a direct proof that $f'(z) = -\frac{1}{z^2}$.

2.12. Prove Proposition 2.6.

2.13. Prove Proposition 2.10.

2.14. Find the derivative of the function $T(z) := \frac{az+b}{cz+d}$, where $a, b, c, d \in \mathbb{C}$ with $ad - bc \neq 0$. When is $T'(z) = 0$?

2.15. Prove that if $f(z)$ is given by a polynomial in z then f is entire. What can you say if $f(z)$ is given by a polynomial in $x = \operatorname{Re} z$ and $y = \operatorname{Im} z$?

2.16. Prove or find a counterexample: If u and v are continuous then $f(z) = u(x,y) + i\,v(x,y)$ is continuous; if u and v are differentiable then f is differentiable.

2.17. Where are the following functions differentiable? Where are they holomorphic? Determine their derivatives at points where they are differentiable.

(a) $f(z) = e^{-x} e^{-iy}$

(b) $f(z) = 2x + ixy^2$

(c) $f(z) = x^2 + iy^2$

(d) $f(z) = e^x e^{-iy}$

(e) $f(z) = \cos x \cosh y - i \sin x \sinh y$

(f) $f(z) = \operatorname{Im} z$

(g) $f(z) = |z|^2 = x^2 + y^2$

(h) $f(z) = z \operatorname{Im} z$

(i) $f(z) = \frac{ix+1}{y}$

(j) $f(z) = 4(\operatorname{Re} z)(\operatorname{Im} z) - i(\overline{z})^2$

(k) $f(z) = 2xy - i(x+y)^2$

(l) $f(z) = z^2 - \overline{z}^2$

2.18. Define $f(z) = 0$ if $\operatorname{Re}(z) \cdot \operatorname{Im}(z) = 0$, and $f(z) = 1$ if $\operatorname{Re}(z) \cdot \operatorname{Im}(z) \neq 0$. Show that f satisfies the Cauchy–Riemann equation (2.2) at $z = 0$, yet f is not differentiable at $z = 0$. Why doesn't this contradict Theorem 2.13(b)?

2.19. Prove: If f is holomorphic in the region $G \subseteq \mathbb{C}$ and always real valued, then f is constant in G. (*Hint*: Use the Cauchy–Riemann equations (2.3) to show that $f' = 0$.)

2.20. Prove: If $f(z)$ and $\overline{f(z)}$ are both holomorphic in the region $G \subseteq \mathbb{C}$ then $f(z)$ is constant in G.

2.21. Suppose f is entire and can be written as $f(z) = v(x) + i\,u(y)$, that is, the real part of f depends only on $x = \operatorname{Re}(z)$ and the imaginary part of f depends only on $y = \operatorname{Im}(z)$. Prove that $f(z) = az + b$ for some $a \in \mathbb{R}$ and $b \in \mathbb{C}$.

2.22. Suppose f is entire, with real and imaginary parts u and v satisfying $u(z)v(z) = 3$ for all z. Show that f is constant.

2.23. Prove that the Cauchy–Riemann equations take on the following form in polar coordinates:
$$\frac{\partial u}{\partial r} = \frac{1}{r}\frac{\partial v}{\partial \varphi} \quad \text{and} \quad \frac{1}{r}\frac{\partial u}{\partial \varphi} = -\frac{\partial v}{\partial r}.$$

2.24. For each of the following functions u, find a function v such that $u + iv$ is holomorphic in some region. Maximize that region.

(a) $u(x, y) = x^2 - y^2$

(b) $u(x, y) = \cosh(y)\sin(x)$

(c) $u(x, y) = 2x^2 + x + 1 - 2y^2$

(d) $u(x, y) = \frac{x}{x^2+y^2}$

2.25. Is $u(x, y) = \frac{x}{x^2+y^2}$ harmonic on $\mathbb{C}$? What about $u(x, y) = \frac{x^2}{x^2+y^2}$?

2.26. Consider the general real homogeneous quadratic function $u(x, y) = ax^2 + bxy + cy^2$, where a, b and c are real constants.

(a) Show that u is harmonic if and only if $a = -c$.

(b) If u is harmonic then show that it is the real part of a function of the form $f(z) = Az^2$ for some $A \in \mathbb{C}$. Give a formula for A in terms of a, b and c.

2.27. Re-prove Proposition 2.10 by using the formula for f' given in Theorem 2.13.

2.28. Prove that, If $G \subseteq \mathbb{C}$ is a region and $f : G \to \mathbb{C}$ is a complex-valued function with $f''(z)$ defined and equal to 0 for all $z \in G$, then $f(z) = az + b$ for some $a, b \in \mathbb{C}$. (*Hint*: Use Theorem 2.17 to show that $f'(z) = a$, and then use Theorem 2.17 again for the function $f(z) - az$.)

Chapter 3

Examples of Functions

To many, mathematics is a collection of theorems. For me, mathematics is a collection of examples; a theorem is a statement about a collection of examples and the purpose of proving theorems is to classify and explain the examples...
John B. Conway

In this chapter we develop a toolkit of complex functions. Our ingredients are familiar from calculus: linear functions, exponentials and logarithms, and trigonometric functions. Yet, when we move these functions into the complex world, they take on—at times drastically different—new features.

3.1 Möbius Transformations

The first class of functions that we will discuss in some detail are built from linear polynomials.

Definition. A *linear fractional transformation* is a function of the form

$$f(z) = \frac{az+b}{cz+d}$$

where $a, b, c, d \in \mathbb{C}$. If $ad - bc \neq 0$ then f is called a *Möbius*[1] *transformation*.

Exercise 2.15 said that any polynomial in z is an entire function, and so the linear fractional transformation $f(z) = \frac{az+b}{cz+d}$ is holomorphic in $\mathbb{C} \setminus \{-\frac{d}{c}\}$, unless $c = 0$ (in which case f is entire). If $c \neq 0$ then $\frac{az+b}{cz+d} = \frac{a}{c}$ implies $ad - bc = 0$, which means that a Möbius transformation $f(z) = \frac{az+b}{cz+d}$ will never take on the value $\frac{a}{c}$. Our first proposition in this chapter says that with these small observations about

---

[1] Named after August Ferdinand Möbius (1790–1868).

the domain and image of a Möbius transformation, we obtain a class of bijections, which are quite special among complex functions.

Proposition 3.1. Let $a, b, c, d \in \mathbb{C}$ with $c \neq 0$. Then $f : \mathbb{C} \setminus \{-\frac{d}{c}\} \to \mathbb{C} \setminus \{\frac{a}{c}\}$ given by $f(z) = \frac{az+b}{cz+d}$ has the inverse function $f^{-1} : \mathbb{C} \setminus \{\frac{a}{c}\} \to \mathbb{C} \setminus \{-\frac{d}{c}\}$ given by

$$f^{-1}(z) = \frac{dz - b}{-cz + a}.$$

We remark that the same formula for $f^{-1}(z)$ works when $c = 0$, except that in this case both domain and image of f are $\mathbb{C}$; see Exercise 3.2. In either case, we note that the inverse of a Möbius transformation is another Möbius transformation.

Example 3.2. Consider the linear fractional transformation $f(z) = \frac{z-1}{iz+i}$. This is a Möbius transformation (check the condition!) with domain $\mathbb{C} \setminus \{-1\}$ whose inverse can be computed via

$$\frac{z-1}{iz+i} = w \quad \Longleftrightarrow \quad z = \frac{iw+1}{-iw+1},$$

so that $f^{-1}(z) = \frac{iz+1}{-iz+1}$, with domain $\mathbb{C} \setminus \{-i\}$. □

Proof of Proposition 3.1. We first prove that f is one-to-one. If $f(z_1) = f(z_2)$, that is,

$$\frac{az_1 + b}{cz_1 + d} = \frac{az_2 + b}{cz_2 + d},$$

then $(az_1 + b)(cz_2 + d) = (az_2 + b)(cz_1 + d)$, which can be rearranged to

$$(ad - bc)(z_1 - z_2) = 0.$$

Since $ad - bc \neq 0$ this implies that $z_1 = z_2$. This shows that f is one-to-one.

Exercise 3.1 verifies that the Möbius transformation $g(z) = \frac{dz-b}{-cz+a}$ is the inverse of f, and by what we have just proved, g is also one-to-one. But this implies that $f : \mathbb{C} \setminus \{-\frac{d}{c}\} \to \mathbb{C} \setminus \{\frac{a}{c}\}$ is onto. □

We remark that Möbius transformations provide an immediate application of Proposition 2.11, as the derivative of $f(z) = \frac{az+b}{cz+d}$ is

$$f'(z) = \frac{a(cz+d) - c(az+b)}{(cz+d)^2} = \frac{ad - bc}{(cz+d)^2}$$

and thus never zero. Proposition 2.11 implies that Möbius transformations are conformal, that is, they preserve angles.

Möbius transformations have even more fascinating geometric properties. En route to an example of such, we introduce some terminology. Special cases of Möbius transformations are *translations* $f(z) = z + b$, *dilations* $f(z) = az$, and *inversion* $f(z) = \frac{1}{z}$. The next result says that if we understand those three special Möbius transformations, we understand them all.

Proposition 3.3. Suppose $f(z) = \frac{az+b}{cz+d}$ is a linear fractional transformation. If $c = 0$ then
$$f(z) = \frac{a}{d} z + \frac{b}{d},$$
and if $c \neq 0$ then
$$f(z) = \frac{bc - ad}{c^2} \frac{1}{z + \frac{d}{c}} + \frac{a}{c}.$$
In particular, every linear fractional transformation is a composition of translations, dilations, and inversions.

Proof. Simplify. □

Theorem 3.4. Möbius transformations map circles and lines into circles and lines.

Example 3.5. Continuing Example 3.2, consider again $f(z) = \frac{z-1}{iz+i}$. For $\varphi \in \mathbb{R}$,
$$f(e^{i\varphi}) = \frac{e^{i\varphi} - 1}{i e^{i\varphi} + i} = \frac{(e^{i\varphi} - 1)(e^{-i\varphi} + 1)}{i|e^{i\varphi} + 1|^2}$$
$$= \frac{e^{i\varphi} - e^{-i\varphi}}{i|e^{i\varphi} + 1|^2} = \frac{2\operatorname{Im}(e^{i\varphi})}{|e^{i\varphi} + 1|^2} = \frac{2\sin\varphi}{|e^{i\varphi} + 1|^2},$$
which is a real number. Thus Theorem 3.4 implies that f maps the unit circle to the real line. □

Proof of Theorem 3.4. Translations and dilations certainly map circles and lines into circles and lines, so by Proposition 3.3, we only have to prove the statement of the theorem for the inversion $f(z) = \frac{1}{z}$.

The equation for a circle centered at $x_0 + i y_0$ with radius r is $(x - x_0)^2 + (y - y_0)^2 = r^2$, which we can transform to
$$\alpha(x^2 + y^2) + \beta x + \gamma y + \delta = 0 \tag{3.1}$$

for some real numbers α, β, γ, and δ that satisfy $\beta^2 + \gamma^2 > 4\alpha\delta$ (Exercise 3.3). The form (3.1) is more convenient for us, because it includes the possibility that the equation describes a line (precisely when $\alpha = 0$).

Suppose $z = x + iy$ satisfies (3.1); we need to prove that $u + iv := \frac{1}{z}$ satisfies a similar equation. Since

$$u + iv = \frac{x - iy}{x^2 + y^2},$$

we can rewrite (3.1) as

$$\begin{aligned} 0 &= \alpha + \beta \frac{x}{x^2 + y^2} + \gamma \frac{y}{x^2 + y^2} + \frac{\delta}{x^2 + y^2} \\ &= \alpha + \beta u - \gamma v + \delta(u^2 + v^2). \end{aligned} \tag{3.2}$$

But this equation, in conjunction with Exercise 3.3, says that $u + iv$ lies on a circle or line. □

3.2 Infinity and the Cross Ratio

In the context of Möbius transformations, it is useful to introduce a formal way of saying that a function f "blows up" in absolute value, and this gives rise to a notion of infinity.

Definition. Suppose $f : G \to \mathbb{C}$.

(a) $\lim_{z \to z_0} f(z) = \infty$ means that for every $M > 0$ we can find $\delta > 0$ so that, for all $z \in G$ satisfying $0 < |z - z_0| < \delta$, we have $|f(z)| > M$.

(b) $\lim_{z \to \infty} f(z) = L$ means that for every $\varepsilon > 0$ we can find $N > 0$ so that, for all $z \in G$ satisfying $|z| > N$, we have $|f(z) - L| < \varepsilon$.

(c) $\lim_{z \to \infty} f(z) = \infty$ means that for every $M > 0$ we can find $N > 0$ so that, for all $z \in G$ satisfying $|z| > N$, we have $|f(z)| > M$.

In the first definition we require that z_0 be an accumulation point of G while in the second and third we require that ∞ be an "extended accumulation point" of G, in the sense that for every $B > 0$ there is some $z \in G$ with $|z| > B$.

As in Section 2.1, the limit, in any of these senses, is unique if it exists.

Example 3.6. We claim that $\lim_{z \to 0} \frac{1}{z^2} = \infty$: Given $M > 0$, let $\delta := \frac{1}{\sqrt{M}}$. Then $0 < |z| < \delta$ implies
$$|f(z)| = \left|\frac{1}{z^2}\right| > \frac{1}{\delta^2} = M. \qquad \square$$

Example 3.7. Let $f(z) = \frac{az+b}{cz+d}$ be a Möbius transformation with $c \neq 0$. Then $\lim_{z \to \infty} f(z) = \frac{a}{c}$.

To simplify the notation, introduce the constant $L := |ad - bc|$. Given $\varepsilon > 0$, let $N := \frac{L}{|c|^2 \varepsilon} + \left|\frac{d}{c}\right|$. Then $|z| > N$ implies, with the reverse triangle inequality (Corollary 1.7((b))), that

$$|cz + d| \geq ||c||z| - |d|| \geq |c||z| - |d| > \frac{L}{|c|\varepsilon}$$

and so

$$\left|f(z) - \frac{a}{c}\right| = \left|\frac{c(az+b) - a(cz+d)}{c(cz+d)}\right| = \frac{L}{|c||cz+d|} < \varepsilon. \qquad \square$$

We stress that ∞ is not a number in $\mathbb{C}$, just as $\pm\infty$ are not numbers in $\mathbb{R}$. However, we can *extend* $\mathbb{C}$ by adding on ∞, if we are careful. We do so by realizing that we are always talking about a limit when handling ∞. It turns out (Exercise 3.11) that the usual limit rules behave well when we mix complex numbers and ∞. For example, if $\lim_{z \to z_0} f(z) = \infty$ and $\lim_{z \to z_0} g(z) = a$ is finite then the usual *limit of sum = sum of limits* rule still holds and gives $\lim_{z \to z_0}(f(z) + g(z)) = \infty$. The following definition captures the philosophy of this paragraph.

Definition. The *extended complex plane* is the set $\hat{\mathbb{C}} := \mathbb{C} \cup \{\infty\}$, together with the following algebraic properties: For any $a \in \mathbb{C}$,

(a) $\infty + a = a + \infty = \infty$ (d) $\frac{a}{\infty} = 0$

(b) if $a \neq 0$ then $\infty \cdot a = a \cdot \infty = \infty$

(c) $\infty \cdot \infty = \infty$ (e) if $a \neq 0$ then $\frac{a}{0} = \infty$.

The extended complex plane is also called the *Riemann sphere* or the *complex projective line*, denoted $\mathbb{CP}^1$.

If a calculation involving ∞ is not covered by the rules above then we must investigate the limit more carefully. For example, it may seem strange that $\infty + \infty$

is not defined, but if we take the limit of $z + (-z) = 0$ as $z \to \infty$ we will get 0, even though the individual limits of z and $-z$ are both ∞.

Now we reconsider Möbius transformations with ∞ in mind. For example, $f(z) = \frac{1}{z}$ is now defined for $z = 0$ and $z = \infty$, with $f(0) = \infty$ and $f(\infty) = 0$, and so we might argue the proper domain for $f(z)$ is actually $\hat{\mathbb{C}}$. Let's consider the other basic types of Möbius transformations. A translation $f(z) = z + b$ is now defined for $z = \infty$, with $f(\infty) = \infty + b = \infty$, and a dilation $f(z) = az$ (with $a \neq 0$) is also defined for $z = \infty$, with $f(\infty) = a \cdot \infty = \infty$. Since every Möbius transformation can be expressed as a composition of translations, dilations and inversions (Proposition 3.3), we see that every Möbius transformation may be interpreted as a transformation of $\hat{\mathbb{C}}$ onto $\hat{\mathbb{C}}$. This general case is summarized in the following extension of Proposition 3.1.

Corollary 3.8. Suppose $ad - bc \neq 0$ and $c \neq 0$, and consider $f : \hat{\mathbb{C}} \to \hat{\mathbb{C}}$ defined through

$$f(z) := \begin{cases} \frac{az+b}{cz+d} & \text{if } z \in \mathbb{C} \setminus \{-\frac{d}{c}\}, \\ \infty & \text{if } z = -\frac{d}{c}, \\ \frac{a}{c} & \text{if } z = \infty. \end{cases}$$

Then f is a bijection.

This corollary also holds for $c = 0$ if we then define $f(\infty) = \infty$.

Example 3.9. Continuing Examples 3.2 and 3.5, consider once more the Möbius transformation $f(z) = \frac{z-1}{iz+i}$. With the definitions $f(-1) = \infty$ and $f(\infty) = -i$, we can extend f to a function $\hat{\mathbb{C}} \to \hat{\mathbb{C}}$. □

With ∞ on our mind we can also add some insight to Theorem 3.4. We recall that in Example 3.5, we proved that $f(z) = \frac{z-1}{iz+i}$ maps the unit circle to the real line. Essentially the same proof shows that, more generally, any circle passing through -1 gets mapped to a line (see Exercise 3.4). The original domain of f was $\mathbb{C} \setminus \{-1\}$, so the point $z = -1$ must be excluded from these circles. However, by thinking of f as function from $\hat{\mathbb{C}}$ to $\hat{\mathbb{C}}$, we can put $z = -1$ back into the picture, and so f transforms circles that pass through -1 to straight lines *plus* ∞. If we remember that ∞ corresponds to being arbitrarily far away from any point in $\mathbb{C}$, we can visualize a line plus ∞ as a circle passing through ∞. If we make this a definition, then Theorem 3.4 can be expressed as: *any Möbius transformation of $\hat{\mathbb{C}}$ transforms circles to circles*.

We can take this remark a step further, based on the idea that three distinct points in $\hat{\mathbb{C}}$ determine a unique circle passing through them: If the three points are in $\mathbb{C}$ and nonlinear, this fact comes straight from Euclidean geometry; if the three points are on a straight line or if one of the points is ∞, then the circle is a straight line plus ∞.

Example 3.10. The Möbius transformation $f : \hat{\mathbb{C}} \to \hat{\mathbb{C}}$ given by $f(z) = \frac{z-1}{iz+i}$ maps

$$1 \mapsto 0, \qquad i \mapsto 1, \qquad \text{and} \qquad -1 \mapsto \infty.$$

The points 1, i, and -1 uniquely determine the unit circle and the points 0, 1, and ∞ uniquely determine the real line, viewed as a circle in $\hat{\mathbb{C}}$. Thus Corollary 3.8 implies that f maps the unit circle to $\mathbb{R}$, which we already concluded in Example 3.5. □

Conversely, if we know where three distinct points in $\hat{\mathbb{C}}$ are transformed by a Möbius transformation then we should be able to figure out everything about the transformation. There is a computational device that makes this easier to see.

Definition. If z, z_1, z_2, and z_3 are any four points in $\hat{\mathbb{C}}$ with z_1, z_2, and z_3 distinct, then their *cross ratio* is defined as

$$[z, z_1, z_2, z_3] := \frac{(z-z_1)(z_2-z_3)}{(z-z_3)(z_2-z_1)}.$$

This includes the implicit definitions $[z_3, z_1, z_2, z_3] = \infty$ and, if one of z, z_1, z_2, or z_3 is ∞, then the two terms containing ∞ are canceled; e.g., $[z, \infty, z_2, z_3] = \frac{z_2-z_3}{z-z_3}$.

Example 3.11. Our running example $f(z) = \frac{z-1}{iz+i}$ can be written as $f(z) = [z, 1, i, -1]$. □

Proposition 3.12. The function $f : \hat{\mathbb{C}} \to \hat{\mathbb{C}}$ defined by $f(z) = [z, z_1, z_2, z_3]$ is a Möbius transformation that satisfies

$$f(z_1) = 0, \qquad f(z_2) = 1, \qquad f(z_3) = \infty.$$

Moreover, if g is any Möbius transformation with $g(z_1) = 0$, $g(z_2) = 1$, and $g(z_3) = \infty$, then f and g are identical.

Proof. Most of this follows from our definition of ∞, but we need to prove the uniqueness statement. By Proposition 3.1, the inverse f^{-1} is a Möbius transformation and, by Exercise 3.10, the composition $h := g \circ f^{-1}$ is again a Möbius transformation. Note that $h(0) = g(f^{-1}(0)) = g(z_1) = 0$ and, similarly, $h(1) = 1$ and $h(\infty) = \infty$. If we write $h(z) = \frac{az+b}{cz+d}$ then

$$0 = h(0) = \frac{b}{d} \implies b = 0$$

$$\infty = h(\infty) = \frac{a}{c} \implies c = 0$$

$$1 = h(1) = \frac{a+b}{c+d} = \frac{a+0}{0+d} = \frac{a}{d} \implies a = d$$

and so

$$h(z) = \frac{az+b}{cz+d} = \frac{az+0}{0+d} = \frac{a}{d} z = z,$$

the identity function. But since $h = g \circ f^{-1}$, this means that f and g are identical. □

So if we want to map three given points of $\hat{\mathbb{C}}$ to 0, 1 and ∞ by a Möbius transformation, then the cross ratio gives us the only way to do it. What if we have three points z_1, z_2 and z_3 and we want to map them to three other points w_1, w_2 and w_3?

Corollary 3.13. Suppose z_1, z_2 and z_3 are distinct points in $\hat{\mathbb{C}}$ and w_1, w_2 and w_3 are distinct points in $\hat{\mathbb{C}}$. Then there is a unique Möbius transformation h satisfying $h(z_1) = w_1$, $h(z_2) = w_2$, and $h(z_3) = w_3$.

Proof. Let $h = g^{-1} \circ f$ where $f(z) = [z, z_1, z_2, z_3]$ and $g(w) = [w, w_1, w_2, w_3]$. Uniqueness follows as in the proof of Proposition 3.12. □

This theorem gives an explicit way to determine h from the points z_j and w_j but, in practice, it is often easier to determine h directly from the conditions $f(z_j) = w_j$ (by solving for a, b, c and d).

3.3 Stereographic Projection

The addition of ∞ to the complex plane $\mathbb{C}$ gives the plane a useful structure. This structure is revealed by a famous function called *stereographic projection*, which gives us a way of visualizing the extended complex plane—that is, with the point at infinity—in $\mathbb{R}^3$, as the unit sphere. It also provides a way of seeing that a line in the extended complex plane really is a circle, and of visualizing Möbius functions.

To begin, we think of $\mathbb{C}$ as the (x,y)-plane in $\mathbb{R}^3$, that is, $\mathbb{C} = \{(x,y,0) \in \mathbb{R}^3\}$. To describe stereographic projection, we will be less concerned with actual complex numbers $x+iy$ and more concerned with their coordinates. Consider the *unit sphere*

$$\mathbb{S}^2 := \{(x,y,z) \in \mathbb{R}^3 : x^2 + y^2 + z^2 = 1\}.$$

The sphere and the complex plane intersect in the set $\{(x,y,0) : x^2 + y^2 = 1\}$, which corresponds to the equator on the sphere and the unit circle on the complex plane, as depicted in Figure 3.1. Let $N := (0,0,1)$, the *north pole* of $\mathbb{S}^2$, and let $S := (0,0,-1)$, the *south pole*.

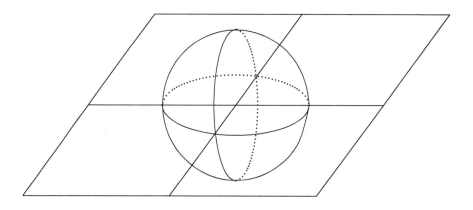

Figure 3.1: Setting up stereographic projection.

Definition. The *stereographic projection of* $\mathbb{S}^2$ *to* $\hat{\mathbb{C}}$ *from* N is the map $\varphi \colon \mathbb{S}^2 \to \hat{\mathbb{C}}$ defined as follows. For any point $P \in \mathbb{S}^2 \setminus \{N\}$, as the z-coordinate of P is strictly less than 1, the line through N and P intersects $\mathbb{C}$ in exactly one point Q. Define $\varphi(P) := Q$. We also declare that $\varphi(N) := \infty$.

Proposition 3.14. The map φ is given by

$$\varphi(x, y, z) = \begin{cases} \left(\frac{x}{1-z}, \frac{y}{1-z}, 0\right) & \text{if } z \neq 1, \\ \infty & \text{if } z = 1. \end{cases}$$

It is bijective, with inverse map

$$\varphi^{-1}(p,q,0) = \left(\frac{2p}{p^2+q^2+1}, \frac{2q}{p^2+q^2+1}, \frac{p^2+q^2-1}{p^2+q^2+1}\right)$$

and $\varphi^{-1}(\infty) = (0,0,1)$.

Proof. Given $P = (x,y,z) \in \mathbb{S}^2 \setminus \{N\}$, the straight line through N and P is parametrized by

$$r(t) = N + t(P-N) = (0,0,1) + t[(x,y,z)-(0,0,1)] = (tx, ty, 1+t(z-1))$$

where $t \in \mathbb{R}$. When $r(t)$ hits $\mathbb{C}$, the third coordinate is 0, so it must be that $t = \frac{1}{1-z}$. Plugging this value of t into the formula for r yields φ as stated.

To see the formula for the inverse map φ^{-1}, we begin with a point $Q = (p,q,0) \in \mathbb{C}$ and solve for a point $P = (x,y,z) \in \mathbb{S}^2$ so that $\varphi(P) = Q$. The point P satisfies the equation $x^2 + y^2 + z^2 = 1$. The equation $\varphi(P) = Q$ tells us that $\frac{x}{1-z} = p$ and $\frac{y}{1-z} = q$. Thus, we solve three equations for three unknowns. The latter two equations yield

$$p^2 + q^2 = \frac{x^2+y^2}{(1-z)^2} = \frac{1-z^2}{(1-z)^2} = \frac{1+z}{1-z}.$$

Solving $p^2 + q^2 = \frac{1+z}{1-z}$ for z and then plugging this into the identities $x = p(1-z)$ and $y = q(1-z)$ proves the desired formula. It is easy to check that $\varphi \circ \varphi^{-1}$ and $\varphi^{-1} \circ \varphi$ are both the identity map; see Exercise 3.25. □

Theorem 3.15. *The stereographic projection φ takes the set of circles in $\mathbb{S}^2$ bijectively to the set of circles in $\hat{\mathbb{C}}$, where for a circle $\gamma \subset \mathbb{S}^2$ we have $\infty \in \varphi(\gamma)$ (that is, $\varphi(\gamma)$ is a line in $\mathbb{C}$) if and only if $N \in \gamma$.*

Proof. A circle in $\mathbb{S}^2$ is the intersection of $\mathbb{S}^2$ with some plane H. If (x_0, y_0, z_0) is a normal vector to H, then there is a unique real number k so that H is given by

$$\begin{aligned} H &= \{(x,y,z) \in \mathbb{R}^3 : (x,y,z) \cdot (x_0, y_0, z_0) = k\} \\ &= \{(x,y,z) \in \mathbb{R}^3 : xx_0 + yy_0 + zz_0 = k\}. \end{aligned}$$

By possibly changing k, we may assume that $(x_0, y_0, z_0) \in \mathbb{S}^2$. We may also assume that $0 \leq k \leq 1$, since if $k < 0$ we can replace (x_0, y_0, z_0) with $(-x_0, -y_0, -z_0)$, and if $k > 1$ then $H \cap \mathbb{S}^2 = \emptyset$.

Consider the circle of intersection $H \cap \mathbb{S}^2$. A point $(p, q, 0)$ in the complex plane lies on the image of this circle under φ if and only if $\varphi^{-1}(p, q, 0)$ satisfies the defining equation for H. Using the equations from Proposition 3.14 for $\varphi^{-1}(p, q, 0)$, we see that
$$(z_0 - k)p^2 + (2x_0)p + (z_0 - k)q^2 + (2y_0)q = z_0 + k.$$

If $z_0 - k = 0$, this is a straight line in the (p, q)-plane. Moreover, every line in the (p, q)-plane can be obtained in this way. Note that $z_0 = k$ if and only if $N \in H$, which is if and only if the image under φ is a straight line.

If $z_0 - k \neq 0$, then completing the square yields
$$\left(p + \frac{x_0}{z_0 - k}\right)^2 + \left(q + \frac{y_0}{z_0 - k}\right)^2 = \frac{1 - k^2}{(z_0 - k)^2}.$$

Depending on whether the right hand side of this equation is positive, 0, or negative, this is the equation of a circle, point, or the empty set in the (p, q)-plane, respectively. These three cases happen when $k < 1$, $k = 1$, and $k > 1$, respectively. Only the first case corresponds to a circle in $\mathbb{S}^2$. Exercise 3.28 verifies that every circle in the (p, q)-plane arises in this manner. □

We can now think of the extended complex plane as a sphere in $\mathbb{R}^3$, the aforementioned *Riemann sphere*.

It is particularly nice to think about the basic Möbius transformations via their effect on the Riemann sphere. We will describe inversion. It is worth thinking about, though beyond the scope of this book, how other Möbius functions behave. For instance, a rotation $f(z) = e^{i\theta} z$, composed with φ^{-1}, can be seen to be a rotation of $\mathbb{S}^2$. We encourage you to verify this and consider the harder problems of visualizing a real dilation, $f(z) = rz$, or a translation, $f(z) = z + b$. We give the hint that a real dilation is in some sense dual to a rotation, in that each moves points along perpendicular sets of circles. Translations can also be visualized via how they move points along sets of circles.

We now use stereographic projection to take another look at $f(z) = \frac{1}{z}$. We want to know what this function does to the sphere $\mathbb{S}^2$. We will take a point $(x, y, z) \in \mathbb{S}^2$, project it to the plane by the stereographic projection φ, apply f to the point that results, and then pull this point back to $\mathbb{S}^2$ by φ^{-1}.

We know $\varphi(x,y,z) = (\frac{x}{1-z}, \frac{y}{1-z}, 0)$ which we now regard as the complex number

$$p + iq = \frac{x}{1-z} + i\frac{y}{1-z}.$$

We know from a previous calculation that $p^2 + q^2 = \frac{1+z}{1-z}$, which gives $x^2 + y^2 = (1+z)(1-z)$. Thus

$$f\left(\frac{x}{1-z} + i\frac{y}{1-z}\right) = \frac{1-z}{x+iy} = \frac{(1-z)(x-iy)}{x^2+y^2} = \frac{x}{1+z} - i\frac{y}{1+z}.$$

Rather than plug this result into the formulas for φ^{-1}, we can just ask what triple of numbers will be mapped to this particular pair using the formulas $\varphi(x,y,z) = (\frac{x}{1-z}, \frac{y}{1-z}, 0)$. The answer is $(x, -y, -z)$.

Thus we have shown that the effect of $f(z) = \frac{1}{z}$ on $\mathbb{S}^2$ is to take (x,y,z) to $(x,-y,-z)$. This is a rotation around the x-axis by 180 degrees.

We now have a second argument that $f(z) = \frac{1}{z}$ takes circles and lines to circles and lines. A circle or line in $\mathbb{C}$ is taken to a circle on $\mathbb{S}^2$ by φ^{-1}. Then $f(z) = \frac{1}{z}$ rotates the sphere which certainly takes circles to circles. Now φ takes circles back to circles and lines. We can also say that the circles that go to lines under $f(z) = \frac{1}{z}$ are the circles through 0, because 0 is mapped to $(0,0,-1)$ under φ^{-1}, and so a circle through 0 in $\mathbb{C}$ goes to a circle through the south pole in $\mathbb{S}^2$. Now 180-degree rotation about the x-axis takes the south pole to the north pole, and our circle is now passing through N. But we know that φ will take this circle to a line in $\mathbb{C}$.

We end by mentioning that there is, in fact, a way of putting the complex metric on $\mathbb{S}^2$. It is certainly not the (finite) distance function induced by $\mathbb{R}^3$. Indeed, the origin in the complex plane corresponds to the south pole of $\mathbb{S}^2$. We have to be able to get arbitrarily far away from the origin in $\mathbb{C}$, so the complex distance function has to increase greatly with the z coordinate. The closer points are to the north pole N (corresponding to ∞ in $\hat{\mathbb{C}}$), the *larger* their distance to the origin, and to each other! In this light, a 'line' in the Riemann sphere $\mathbb{S}^2$ corresponds to a circle in $\mathbb{S}^2$ through N. In the regular sphere, the circle has finite length, but as a line on the Riemann sphere with the complex metric, it has infinite length.

3.4 Exponential and Trigonometric Functions

To define the complex exponential function, we once more borrow concepts from calculus, namely the real exponential function[2] and the real sine and cosine, and we finally make sense of the notation $e^{it} = \cos t + i \sin t$.

Definition. The *(complex) exponential function* is $\exp : \mathbb{C} \to \mathbb{C}$, defined for $z = x + iy$ as
$$\exp(z) := e^x (\cos y + i \sin y) = e^x e^{iy}.$$

This definition seems a bit arbitrary. Our first justification is that all exponential rules that we are used to from real numbers carry over to the complex case. They mainly follow from Proposition 1.3 and are collected in the following.

Proposition 3.16. *For all* $z, z_1, z_2 \in \mathbb{C}$,

(a) $\exp(z_1) \exp(z_2) = \exp(z_1 + z_2)$

(b) $\frac{1}{\exp(z)} = \exp(-z)$

(c) $\exp(z + 2\pi i) = \exp(z)$

(d) $|\exp(z)| = \exp(\operatorname{Re} z)$

(e) $\exp(z) \neq 0$

(f) $\frac{d}{dz} \exp(z) = \exp(z)$.

Identity ((c)) is very special and has no counterpart for the real exponential function. It says that the complex exponential function is *periodic* with period $2\pi i$. This has many interesting consequences; one that may not seem too pleasant at first sight is the fact that the complex exponential function is not one-to-one.

Identity ((f)) is not only remarkable, but we invite you to meditate on its proof below; it gives a strong indication that our definition of exp is reasonable. We also note that ((f)) says that exp is entire.

We leave the proof of Proposition 3.16 as Exercise 3.34 but give one sample.

Proof of ((f))*.* The partial derivatives of $f(z) = \exp(z)$ are

$$\frac{\partial f}{\partial x} = e^x (\cos y + i \sin y) \quad \text{and} \quad \frac{\partial f}{\partial y} = e^x (-\sin y + i \cos y).$$

---

[2] How to define the real exponential function is a nontrivial question. Our preferred way to do this is through a power series: $e^x = \sum_{k \geq 0} \frac{1}{k!} x^k$. In light of this definition, you might think we should have simply defined the complex exponential function through a complex power series. In fact, this is possible (and an elegant definition); however, one of the promises of this book is to introduce complex power series as late as possible. We agree with those readers who think that we are cheating at this point, as we borrow the concept of a (real) power series to define the real exponential function.

They are continuous in $\mathbb{C}$ and satisfy the Cauchy–Riemann equation (2.2):

$$\frac{\partial f}{\partial x}(z) = -i\frac{\partial f}{\partial y}(z)$$

for all $z \in \mathbb{C}$. Thus Theorem 2.13 says that $f(z) = \exp(z)$ is entire with derivative

$$f'(z) = \frac{\partial f}{\partial x}(z) = \exp(z). \qquad \square$$

We should make sure that the complex exponential function specializes to the real exponential function for real arguments: indeed, if $z = x \in \mathbb{R}$ then

$$\exp(x) = e^x(\cos 0 + i \sin 0) = e^x.$$

The trigonometric functions—sine, cosine, tangent, cotangent, etc.—also have complex analogues; however, they do not play the same prominent role as in the real case. In fact, we can define them as merely being special combinations of the exponential function.

Definition. The *(complex) sine* and *cosine* are defined as

$$\sin z := \tfrac{1}{2i}\left(\exp(iz) - \exp(-iz)\right) \qquad \text{and} \qquad \cos z := \tfrac{1}{2}\left(\exp(iz) + \exp(-iz)\right),$$

respectively. The *tangent* and *cotangent* are defined as

$$\tan z := \frac{\sin z}{\cos z} = -i\,\frac{\exp(2iz) - 1}{\exp(2iz) + 1}$$

and

$$\cot z := \frac{\cos z}{\sin z} = i\,\frac{\exp(2iz) + 1}{\exp(2iz) - 1},$$

respectively.

Note that to write tangent and cotangent in terms of the exponential function, we used the fact that $\exp(z)\exp(-z) = \exp(0) = 1$. Because $\exp$ is entire, so are $\sin$ and $\cos$.

As with the exponential function, we should make sure that we're not redefining the real sine and cosine: if $z = x \in \mathbb{R}$ then

$$\begin{aligned}\sin z &= \tfrac{1}{2i}(\exp(ix) - \exp(-ix)) \\ &= \tfrac{1}{2i}(\cos x + i \sin x - \cos(-x) - i \sin(-x)) = \sin x.\end{aligned}$$

A similar calculation holds for the cosine. Not too surprisingly, the following properties follow mostly from Proposition 3.16.

Proposition 3.17. For all $z, z_1, z_2 \in \mathbb{C}$,

$$\sin(-z) = -\sin z \qquad\qquad \cos(-z) = \cos z$$
$$\sin(z + 2\pi) = \sin z \qquad\qquad \cos(z + 2\pi) = \cos z$$
$$\tan(z + \pi) = \tan z \qquad\qquad \cot(z + \pi) = \cot z$$
$$\sin(z + \tfrac{\pi}{2}) = \cos z \qquad\qquad \cos(z + \tfrac{\pi}{2}) = -\sin z$$
$$\sin(z_1 + z_2) = \sin z_1 \cos z_2 + \cos z_1 \sin z_2$$
$$\cos(z_1 + z_2) = \cos z_1 \cos z_2 - \sin z_1 \sin z_2$$
$$\cos^2 z + \sin^2 z = 1 \qquad\qquad \cos^2 z - \sin^2 z = \cos(2z)$$
$$\frac{d}{dz}\sin z = \cos z \qquad\qquad \frac{d}{dz}\cos z = -\sin z.$$

Finally, one word of caution: unlike in the real case, the complex sine and cosine are *not* bounded—consider, for example, $\sin(iy)$ as $y \to \pm\infty$.

We end this section with a remark on hyperbolic trig functions. The *hyperbolic sine, cosine, tangent,* and *cotangent* are defined as in the real case:

Definition.

$$\sinh z = \tfrac{1}{2}(\exp(z) - \exp(-z)) \qquad \cosh z = \tfrac{1}{2}(\exp(z) + \exp(-z))$$
$$\tanh z = \frac{\sinh z}{\cosh z} = \frac{\exp(2z) - 1}{\exp(2z) + 1} \qquad \coth z = \frac{\cosh z}{\sinh z} = \frac{\exp(2z) + 1}{\exp(2z) - 1}.$$

As such, they are yet more special combinations of the exponential function. They still satisfy the identities you already know, e.g.,

$$\frac{d}{dz}\sinh z = \cosh z \quad \text{and} \quad \frac{d}{dz}\cosh z = \sinh z.$$

Moreover, they are related to the trigonometric functions via

$$\sinh(iz) = i\sin z \quad \text{and} \quad \cosh(iz) = \cos z.$$

3.5 Logarithms and Complex Exponentials

The complex logarithm is the first function we'll encounter that is of a somewhat tricky nature. It is motivated as an inverse to the exponential function, that is, we're looking for a function Log such that

$$\exp(\text{Log}(z)) = z = \text{Log}(\exp z). \tag{3.3}$$

But because exp is not one-to-one, this is too much to hope for. In fact, given a function Log that satisfies the first equation in (3.3), the function $f(z) = \text{Log}(z) + 2\pi i$ does as well, and so there cannot be an inverse of exp (which would have to be unique). On the other hand, exp becomes one-to-one if we restrict its domain, so there is hope for a logarithm if we're careful about its construction and about its domain.

Definition. Given a region G, any continuous function $\text{Log}: G \to \mathbb{C}$ that satisfies $\exp(\text{Log } z) = z$ is *a branch of the logarithm (on G)*.

To make sure this definition is not vacuous, let's write, as usual, $z = re^{i\varphi}$, and suppose that $\text{Log } z = u(z) + i\,v(z)$. Then for the first equation in (3.3) to hold, we need

$$\exp(\text{Log } z) = e^u e^{iv} = re^{i\varphi} = z,$$

that is, $e^u = r$ and $e^{iv} = e^{i\varphi}$. The latter equation is equivalent to $v = \varphi + 2\pi k$ for some $k \in \mathbb{Z}$, and denoting the natural logarithm of the positive real number x by $\ln(x)$, the former equation is equivalent to $u = \ln|z|$. A reasonable definition of a logarithm function Log would hence be $\text{Log } z = \ln|z| + i\,\text{Arg } z$ where $\text{Arg } z$ gives the argument for the complex number z according to some convention—here is an example:

Definition. Let $\text{Arg } z$ denote the unique argument of $z \neq 0$ that lies in $(-\pi, \pi]$ (the *principal argument of z*). Then the *principal logarithm* is the function $\text{Log}:$

$\mathbb{C} \setminus \{0\} \to \mathbb{C}$ defined through

$$\mathrm{Log}(z) := \ln|z| + i\,\mathrm{Arg}(z).$$

Example 3.18. Here are a few evaluations of Log to illustrate this function:

$$\begin{aligned}
\mathrm{Log}(2) &= \ln(2) + i\,\mathrm{Arg}(2) = \ln(2) \\
\mathrm{Log}(i) &= \ln(1) + i\,\mathrm{Arg}(i) = \frac{\pi i}{2} \\
\mathrm{Log}(-3) &= \ln(3) + i\,\mathrm{Arg}(-3) = \ln(3) + \pi i \\
\mathrm{Log}(1-i) &= \ln(\sqrt{2}) + i\,\mathrm{Arg}(1-i) = \frac{1}{2}\ln(2) - \frac{\pi i}{4}.
\end{aligned}$$

□

The principal logarithm is not continuous on the negative part of the real line, and so Log is a branch of the logarithm on $\mathbb{C} \setminus \mathbb{R}_{\leq 0}$. Any branch of the logarithm on a region G can be similarly extended to a function defined on $\overline{G} \setminus \{0\}$. Furthermore, the evaluation of any branch of the logarithm at a specific z_0 can differ from $\mathrm{Log}(z_0)$ only by a multiple of $2\pi i$; the reason for this is once more the periodicity of the exponential function.

So what about the second equation in (3.3), namely, $\mathrm{Log}(\exp z) = z$? Let's try the principal logarithm: if $z = x + iy$ then

$$\mathrm{Log}(\exp z) = \ln|e^x e^{iy}| + i\,\mathrm{Arg}(e^x e^{iy}) = \ln e^x + i\,\mathrm{Arg}(e^{iy}) = x + i\,\mathrm{Arg}(e^{iy}).$$

The right-hand side is equal to $z = x + iy$ if and only if $y \in (-\pi, \pi]$. Something similar will happen with any other branch $\log$ of the logarithm—there will always be many z's for which $\log(\exp z) \neq z$.

A warning sign pointing in a similar direction concerns the much-cherished real logarithm rule $\ln(xy) = \ln(x) + \ln(y)$; it has no analogue in $\mathbb{C}$. For example,

$$\mathrm{Log}(i) + \mathrm{Log}(i-1) = i\tfrac{\pi}{2} + \ln\sqrt{2} + \tfrac{3\pi i}{4} = \tfrac{1}{2}\ln 2 + \tfrac{5\pi i}{4}$$

but

$$\mathrm{Log}(i(i-1)) = \mathrm{Log}(-1-i) = \tfrac{1}{2}\ln 2 - \tfrac{3\pi i}{4}.$$

The problem is that we need to come up with an argument convention to define a logarithm and then stick to this convention. There is quite a bit of subtlety here;

e.g., the multi-valued map

$$\arg z := \text{all possible arguments of } z$$

gives rise to a multi-valued logarithm via

$$\log z := \ln|z| + i \arg z.$$

Neither arg nor log is a function, yet $\exp(\log z) = z$. We invite you to check this thoroughly; in particular, you should note how the periodicity of the exponential function takes care of the multi-valuedness of log.

To end our discussion of complex logarithms on a happy note, we prove that *any* branch of the logarithm has the *same* derivative; one just has to be cautious with regions of holomorphicity.

Proposition 3.19. If $\mathcal{L}\text{og}$ is a branch of the logarithm on G then $\mathcal{L}\text{og}$ is differentiable on G with

$$\frac{d}{dz} \mathcal{L}\text{og}(z) = \frac{1}{z}.$$

Proof. The idea is to apply Proposition 2.12 to exp and $\mathcal{L}\text{og}$, but we need to be careful about the domains of these functions. Let $H := \{\mathcal{L}\text{og}(z) : z \in G\}$, the image of $\mathcal{L}\text{og}$. We apply Proposition 2.12 with $f : H \to G$ given by $f(z) = \exp(z)$ and $g : G \to H$ given by $g(z) = \mathcal{L}\text{og}(z)$; note that g is continuous, and Exercise 3.48 checks that f and g are inverses of each other. Thus Proposition 2.12 gives

$$\mathcal{L}\text{og}'(z) = \frac{1}{\exp'(\mathcal{L}\text{og} z)} = \frac{1}{\exp(\mathcal{L}\text{og} z)} = \frac{1}{z}. \qquad \square$$

We finish this section by defining complex exponentials.

Definition. Given $a, b \in \mathbb{C}$ with $a \neq 0$, the *principal value of* a^b is defined as

$$a^b := \exp(b \, \text{Log}(a)).$$

Naturally, we can just as well define a^b through a different branch of the logarithm; our convention is that we use the principal value unless otherwise stated. Exercise 3.51 explores what happens when we use the multi-valued log in the definition of a^b.

One last remark: it now makes sense to talk about the function $f(z) = e^z$, where e is *Euler's*[3] *number* and can be defined, for example, as $e = \lim_{n\to\infty} \left(1 + \frac{1}{n}\right)^n$. In calculus we can prove the equivalence of the real exponential function (as given, for example, through a power series) and the function $f(x) = e^x$. With our definition of a^z, we can now make a similar remark about the complex exponential function. Because e is a positive real number and hence $\operatorname{Arg} e = 0$,

$$e^z = \exp(z \operatorname{Log}(e)) = \exp(z(\ln|e| + i \operatorname{Arg}(e))) = \exp(z \ln(e)) = \exp(z).$$

A word of caution: this only works out this nicely because we have carefully defined a^b for complex numbers. Using a different branch of logarithm in the definition of a^b can easily lead to $e^z \neq \exp(z)$.

Exercises

3.1. Show that if $f(z) = \frac{az+b}{cz+d}$ is a Möbius transformation then $f^{-1}(z) = \frac{dz-b}{-cz+a}$.

3.2. Complete the picture painted by Proposition 3.1 by considering Möbius transformations with $c = 0$. That is, show that $f : \mathbb{C} \to \mathbb{C}$ given by $f(z) = \frac{az+b}{d}$ is a bijection, with $f^{-1}(z)$ given by the formula in Proposition 3.1.

3.3. Show that (3.1) is the equation for a circle or line if and only if $\beta^2 + \gamma^2 > 4\alpha\delta$. Conclude that $x + iy$ is a solution to (3.1) if and only if $u + iv$ is a solution to (3.2).

3.4. Extend Example 3.5 by showing that $f(z) = \frac{z-1}{iz+i}$ maps any circle passing through -1 to a line.

3.5. Prove that any Möbius transformation different from the identity map can have at most two fixed points. (A *fixed point* of a function f is a number z such that $f(z) = z$.)

3.6. Prove Proposition 3.3.

---
[3] Named after Leonard Euler (1707–1783). Continuing our footnote on p. 8, we have now honestly established *Euler's formula* $e^{2\pi i} = 1$.

3.7. Show that the Möbius transformation $f(z) = \frac{1+z}{1-z}$ maps the unit circle (minus the point $z = 1$) onto the imaginary axis.

3.8. Suppose that f is holomorphic in the region G and $f(G)$ is a subset of the unit circle. Show that f is constant.

3.9. Fix $a \in \mathbb{C}$ with $|a| < 1$ and consider

$$f_a(z) := \frac{z - a}{1 - \bar{a}z}.$$

(a) Show that $f_a(z)$ is a Möbius transformation.

(b) Show that $f_a^{-1}(z) = f_{-a}(z)$.

(c) Prove that $f_a(z)$ maps the unit disk $D[0, 1]$ to itself in a bijective fashion.

3.10. Suppose

$$A = \begin{bmatrix} a & b \\ c & d \end{bmatrix}$$

is a 2×2 matrix of complex numbers whose determinant $ad - bc$ is nonzero. Then we can define a corresponding Möbius transformation on $\hat{\mathbb{C}}$ by $T_A(z) = \frac{az+b}{cz+d}$. Show that $T_A \circ T_B = T_{A \cdot B}$, where $\circ$ denotes composition and $\cdot$ denotes matrix multiplication.

3.11. Show that our definition of $\hat{\mathbb{C}}$ honors the "finite" limit rules in Proposition 2.4, by proving the following, where $a \in \mathbb{C}$:

(a) If $\lim_{z \to z_0} f(z) = \infty$ and $\lim_{z \to z_0} g(z) = a$ then $\lim_{z \to z_0} (f(z) + g(z)) = \infty$.

(b) If $\lim_{z \to z_0} f(z) = \infty$ and $\lim_{z \to z_0} g(z) = a \neq 0$ then $\lim_{z \to z_0} (f(z) \cdot g(z)) = \infty$.

(c) If $\lim_{z \to z_0} f(z) = \lim_{z \to z_0} g(z) = \infty$ then $\lim_{z \to z_0} (f(z) \cdot g(z)) = \infty$.

(d) If $\lim_{z \to z_0} f(z) = \infty$ and $\lim_{z \to z_0} g(z) = a$ then $\lim_{z \to z_0} \frac{g(z)}{f(z)} = 0$.

(e) If $\lim_{z \to z_0} f(z) = 0$ and $\lim_{z \to z_0} g(z) = a \neq 0$ then $\lim_{z \to z_0} \frac{g(z)}{f(z)} = \infty$.

3.12. Fix $c_0, c_1, \ldots, c_{d-1} \in \mathbb{C}$. Prove that

$$\lim_{z \to \infty} 1 + \frac{c_{d-1}}{z} + \frac{c_{d-2}}{z^2} + \cdots + \frac{c_0}{z^d} = 1.$$

3.13. Let $f(z) = \frac{2z}{z+2}$. Draw two graphs, one showing the following six sets in the z-plane and the other showing their images in the w-plane. Label the sets. (You should only need to calculate the images of $0, \pm 2, \infty$ and $-1-i$; remember that Möbius transformations preserve angles.)

(a) the x-axis plus ∞

(b) the y-axis plus ∞

(c) the line $x = y$ plus ∞

(d) the circle with radius 2 centered at 0

(e) the circle with radius 1 centered at 1

(f) the circle with radius 1 centered at -1

3.14. Find Möbius transformations satisfying each of the following. Write your answers in standard form, as $\frac{az+b}{cz+d}$.

(a) $1 \to 0,\ 2 \to 1,\ 3 \to \infty$

(b) $1 \to 0,\ 1+i \to 1,\ 2 \to \infty$

(c) $0 \to i,\ 1 \to 1,\ \infty \to -i$.

3.15. Using the cross ratio, with different choices of z_k, find two different Möbius transformations that transform $C[1+i, 1]$ onto the real axis plus ∞. In each case, find the image of the center of the circle.

3.16. Let γ be the unit circle. Find a Möbius transformation that transforms γ onto γ and transforms 0 to $\frac{1}{2}$.

3.17. Describe the image of the region under the transformation:

(a) The disk $|z| < 1$ under $w = \frac{iz-i}{z+1}$.

(b) The quadrant $x > 0$, $y > 0$ under $w = \frac{z-i}{z+i}$.

(c) The strip $0 < x < 1$ under $w = \frac{z}{z-1}$.

3.18. Find a Möbius transformation that maps the unit disk to $\{x+iy \in \mathbb{C} : x+y > 0\}$.

3.19. The *Jacobian* of a transformation $u = u(x, y)$, $v = v(x, y)$ is the determinant of the matrix

$$\begin{bmatrix} \frac{\partial u}{\partial x} & \frac{\partial u}{\partial y} \\ \frac{\partial v}{\partial x} & \frac{\partial v}{\partial y} \end{bmatrix}.$$

Show that if $f = u + iv$ is holomorphic then the Jacobian equals $|f'(z)|^2$.

3.20. Find the fixed points in $\hat{\mathbb{C}}$ of $f(z) = \frac{z^2-1}{2z+1}$.

3.21. Find each Möbius transformation f:

(a) f maps $0 \to 1$, $1 \to \infty$, $\infty \to 0$.

(b) f maps $1 \to 1$, $-1 \to i$, $-i \to -1$.

(c) f maps the x-axis to $y = x$, the y-axis to $y = -x$, and the unit circle to itself.

3.22.

(a) Find a Möbius transformation that maps the unit circle to $\{x + iy \in \mathbb{C} : x + y = 0\}$.

(b) Find two Möbius transformations that map the unit disk

$\{z \in \mathbb{C} : |z| < 1\}$ to $\{x+iy \in \mathbb{C} : x+y > 0\}$ and $\{x+iy \in \mathbb{C} : x+y < 0\}$,

respectively.

3.23. Given $a \in \mathbb{R} \setminus \{0\}$, show that the image of the line $y = a$ under inversion is the circle with center $\frac{-i}{2a}$ and radius $\frac{1}{2a}$.

3.24. Suppose z_1, z_2 and z_3 are distinct points in $\hat{\mathbb{C}}$. Show that z is on the circle passing through z_1, z_2 and z_3 if and only if $[z, z_1, z_2, z_3]$ is real or ∞.

3.25. Prove that the stereographic projection of Proposition 3.14 is a bijection by verifying that $\varphi \circ \varphi^{-1}$ and $\varphi^{-1} \circ \varphi$ are the identity map.

3.26. Find the image of the following points under the stereographic projection φ:
$(0,0,-1), (0,0,1), (1,0,0), (0,1,0), (1,1,0)$.

3.27. Consider the plane H determined by $x + y - z = 0$. What is a unit normal vector to H? Compute the image of $H \cap \mathbb{S}^2$ under the stereographic projection φ.

3.28. Prove that every circle in the extended complex plane $\hat{\mathbb{C}}$ is the image of some circle in $\mathbb{S}^2$ under the stereographic projection φ.

3.29. Describe the effect of the basic Möbius transformations rotation, real dilation, and translation on the Riemann sphere. (*Hint*: For the first two, consider all circles in $\mathbb{S}^2$ centered on the NS axis, and all circles through both N and S. For translation, consider two families of circles through N, orthogonal to and perpendicular to the translation.)

3.30. Prove that $\overline{\sin(z)} = \sin(\overline{z})$ and $\overline{\cos(z)} = \cos(\overline{z})$.

3.31. Let $z = x + iy$ and show that
(a) $\sin z = \sin x \cosh y + i \cos x \sinh y$.
(b) $\cos z = \cos x \cosh y - i \sin x \sinh y$.

3.32. Prove that the zeros of $\sin z$ are all real valued. Conclude that they are precisely the integer multiples of π.

3.33. Describe the images of the following sets under the exponential function $\exp(z)$:

(a) the line segment defined by $z = iy$, $0 \leq y \leq 2\pi$

(b) the line segment defined by $z = 1 + iy$, $0 \leq y \leq 2\pi$

(c) the rectangle $\{z = x + iy \in \mathbb{C} : 0 \leq x \leq 1, 0 \leq y \leq 2\pi\}$.

3.34. Prove Proposition 3.16.

3.35. Prove Proposition 3.17.

3.36. Let $z = x + iy$ and show that

(a) $|\sin z|^2 = \sin^2 x + \sinh^2 y = \cosh^2 y - \cos^2 x$

(b) $|\cos z|^2 = \cos^2 x + \sinh^2 y = \cosh^2 y - \sin^2 x$

(c) If $\cos x = 0$ then
$$|\cot z|^2 = \frac{\cosh^2 y - 1}{\cosh^2 y} \leq 1.$$

(d) If $|y| \geq 1$ then
$$|\cot z|^2 \leq \frac{\sinh^2 y + 1}{\sinh^2 y} = 1 + \frac{1}{\sinh^2 y} \leq 1 + \frac{1}{\sinh^2 1} \leq 2.$$

3.37. Show that $\tan(iz) = i \tanh(z)$.

3.38. Draw a picture of the images of vertical lines under the sine function. Do the same for the tangent function.

3.39. Determine the image of the strip $\{z \in \mathbb{C} : -\frac{\pi}{2} < \operatorname{Re} z < \frac{\pi}{2}\}$ under the sine function. (*Hint:* Exercise 3.31 makes it easy to convert parametric equations for horizontal or vertical lines to parametric equations for their images. Note that the equations $x = A \sin t$ and $y = B \cos t$ represent an ellipse and the equations $x = A \cosh t$ and $y = B \sinh t$ represent a hyperbola. Start by finding the images of the boundary lines of the strip, and then find the images of a few horizontal segments and vertical lines in the strip.)

3.40. Find the principal values of

(a) $\text{Log}(2i)$

(b) $(-1)^i$

(c) $\text{Log}(-1+i)$.

3.41. Convert the following expressions to the form $x+iy$. (Reason carefully.)

(a) $e^{i\pi}$

(b) e^π

(c) i^i

(d) $e^{\sin(i)}$

(e) $\exp(\text{Log}(3+4i))$

(f) $(1+i)^{\frac{1}{2}}$

(g) $\sqrt{3}(1-i)$

(h) $\left(\frac{i+1}{\sqrt{2}}\right)^4$.

3.42. Is $\arg(\bar{z}) = -\arg(z)$ true for the multiple-valued argument? What about $\text{Arg}(\bar{z}) = -\text{Arg}(z)$ for the principal branch?

3.43. For the multiple-valued logarithm, is there a difference between the set of all values of $\log(z^2)$ and the set of all values of $2\log z$? (*Hint*: Try some fixed numbers for z.)

3.44. For each of the following functions, determine all complex numbers for which the function is holomorphic. If you run into a logarithm, use the principal value unless otherwise stated.

(a) $\bar{z}^2$

(b) $\frac{\sin z}{z^3+1}$

(c) $\text{log}(z-2i+1)$ where $\text{log}(z) = \ln|z| + i\,\text{Arg}(z)$ with $0 \leq \text{Arg}(z) < 2\pi$

(d) $\exp(\bar{z})$

(e) $(z-3)^i$

(f) i^{z-3}.

3.45. Find all solutions to the following equations:

(a) $\text{Log}(z) = \frac{\pi i}{2}$

(b) $\text{Log}(z) = \frac{3\pi i}{2}$

(c) $\exp(z) = \pi i$

(d) $\sin(z) = \cosh(4)$

(e) $\cos(z) = 0$

(f) $\sinh(z) = 0$

(g) $\overline{\exp(iz)} = \exp(i\bar{z})$

(h) $z^{\frac{1}{2}} = 1 + i$.

3.46. Find the image of the annulus $1 < |z| < e$ under the principal value of the logarithm.

3.47. Use Exercise 2.23 to give an alternative proof that Log is holomorphic in $\mathbb{C} \setminus \mathbb{R}_{\leq 0}$.

3.48. Let log be a branch of the logarithm on G, and let $H := \{\text{log}(z) : z \in G\}$, the image of log. Show that $\text{log} : G \to H$ is a bijection whose inverse map is $f(z) : H \to G$ given by $f(z) = \exp(z)$ (i.e., f is the exponential function restricted to H).

3.49. Show that $|a^z| = a^{\text{Re}\, z}$ if a is a positive real constant.

3.50. Fix $c \in \mathbb{C} \setminus \{0\}$. Find the derivative of $f(z) = z^c$.

3.51. Prove that $\exp(b \log a)$ is single valued if and only if b is an integer. (Note that this means that complex exponentials do not clash with monomials z^n, no matter which branch of the logarithm is used.) What can you say if b is rational?

3.52. Describe the image under exp of the line with equation $y = x$. To do this you should find an equation (at least parametrically) for the image (you can start with the parametric form $x = t, y = t$), plot it reasonably carefully, and explain what happens in the limits as $t \to \infty$ and $t \to -\infty$.

3.53. For this problem, $f(z) = z^2$.

(a) Show that the image under f of a circle centered at the origin is a circle centered at the origin.

(b) Show that the image under f of a ray starting at the origin is a ray starting at the origin.

(c) Let T be the figure formed by the horizontal segment from 0 to 2, the circular arc from 2 to $2i$, and then the vertical segment from $2i$ to 0. Draw T and $f(T)$.

(d) Is the right angle at the origin in part (c) preserved? Is something wrong here? (*Hint*: Use polar coordinates.)

3.54. As in the previous problem, let $f(z) = z^2$. Let Q be the square with vertices at $0, 2, 2+2i$ and $2i$. Draw $f(Q)$ and identify the types of image curves corresponding to the segments from 2 to $2+2i$ and from $2+2i$ to $2i$. They are not parts of either straight lines or circles. (*Hint*: You can write the vertical segment parametrically as $z(t) = 2 + it$. Eliminate the parameter in $u + iv = f(z(t))$ to get a (u, v) equation for the image curve.) Exercises 3.53 and 3.54 are related to the cover picture of this book.

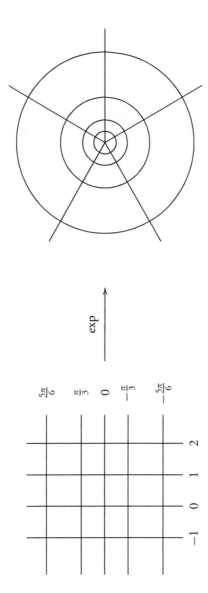

Figure 3.2: Image properties of the exponential function.

Chapter 4

Integration

If things are nice there is probably a good reason why they are nice: and if you do not know at least one reason for this good fortune, then you still have work to do.
Richard Askey

We are now ready to start integrating complex functions—and we will not stop doing so for the remainder of this book: it turns out that complex integration is much richer than real integration (in one variable). The initial reason for this is that we have an extra dimension to play with: the calculus integral $\int_a^b f(x)\,dx$ has a fixed integration path, from a to b along the real line. For complex functions, there are many different ways to go from a to b...

4.1 Definition and Basic Properties

At first sight, complex integration is not really different from real integration. Let $a, b \in \mathbb{R}$ and let $g : [a, b] \to \mathbb{C}$ be continuous. Then we define

$$\int_a^b g(t)\,dt := \int_a^b \operatorname{Re} g(t)\,dt + i \int_a^b \operatorname{Im} g(t)\,dt. \qquad (4.1)$$

This definition is analogous to that of integration of a parametric curve in $\mathbb{R}^2$. For a function that takes complex numbers as arguments, we typically integrate over a path γ (in place of a real interval). If you meditate about the substitution rule for real integrals (Theorem A.6), the following definition, which is based on (4.1), should come as no surprise.

Definition. Suppose γ is a smooth path parametrized by $\gamma(t)$, $a \le t \le b$, and f is a complex function which is continuous on γ. Then we define the *integral of f on γ* as

$$\int_\gamma f = \int_\gamma f(z)\,dz := \int_a^b f(\gamma(t))\gamma'(t)\,dt.$$

DEFINITION AND BASIC PROPERTIES

This definition immediately extends to paths that are *piecewise smooth*: Suppose γ is parametrized by $\gamma(t)$, $a \leq t \leq b$, which is smooth on the intervals $[a,c_1]$, $[c_1,c_2]$, ..., $[c_{n-1},c_n]$, $[c_n,b]$.[1] Then, assuming again that f is continuous on γ, we define

$$\int_\gamma f := \int_a^{c_1} f(\gamma(t))\gamma'(t)\,dt + \int_{c_1}^{c_2} f(\gamma(t))\gamma'(t)\,dt + \cdots + \int_{c_n}^{b} f(\gamma(t))\gamma'(t)\,dt.$$

Example 4.1. To see this definition in action, we compute the integral of the function $f : \mathbb{C} \to \mathbb{C}$ given by $f(z) = \bar{z}^2$ over several paths from 0 to $1+i$.

(a) Let γ be the line segment from 0 to $1+i$. A parametrization of this path is $\gamma(t) = t + it$, $0 \leq t \leq 1$. Here $\gamma'(t) = 1+i$ and $f(\gamma(t)) = (t-it)^2$, and so

$$\int_\gamma f = \int_0^1 (t-it)^2 (1+i)\,dt = (1+i)\int_0^1 (t^2 - 2it^2 - t^2)\,dt$$
$$= -\frac{2i(1+i)}{3} = \frac{2}{3}(1-i).$$

(b) Let γ be the arc of the parabola $y = x^2$ from 0 to $1+i$. A parametrization of this path is $\gamma(t) = t + it^2$, $0 \leq t \leq 1$. Now we have $\gamma'(t) = 1 + 2it$ and

$$f(\gamma(t)) = (t - it)^2 = t^2 - t^4 - 2it^3,$$

whence

$$\int_\gamma f = \int_0^1 (t^2 - t^4 - 2it^3)(1 + 2it)\,dt = \int_0^1 (t^2 + 3t^4 - 2it^5)\,dt$$
$$= \frac{1}{3} + 3\frac{1}{5} - 2i\frac{1}{6} = \frac{14}{15} - \frac{i}{3}.$$

(c) Let γ be the union of the two line segments γ_1 from 0 to 1 and γ_2 from 1 to $1+i$. Parametrizations are $\gamma_1(t) = t$, $0 \leq t \leq 1$, and $\gamma_2(t) = 1+it$, $0 \leq t \leq 1$.

---

[1] Our footnote on p. 14 about the subtlety of the definition of a smooth path applies also here, at the subdivision points c_j. Note that we do *not* require that the left and right derivatives match at these points.

Hence

$$\int_\gamma f = \int_{\gamma_1} f + \int_{\gamma_2} f = \int_0^1 t^2\,dt + \int_0^1 (1-it)^2\,i\,dt$$
$$= \frac{1}{3} + i\int_0^1 (1 - 2it - t^2)\,dt$$
$$= \frac{1}{3} + i\left(1 - 2i\frac{1}{2} - \frac{1}{3}\right) = \frac{4}{3} + \frac{2}{3}i. \qquad \square$$

It is apparent but nevertheless noteworthy that these integrals evaluate to different results; in particular—unlike in calculus—a complex integral does not simply depend on the endpoints of the path of integration.

On the other hand, the complex integral has some standard properties, most of which follow from their real siblings in a straightforward way. Our first observation is that the actual choice of parametrization of γ does not matter. More precisely, if $\gamma(t)$, $a \le t \le b$ and $\sigma(t)$, $c \le t \le d$ are parametrizations of a curve then we say that σ is a *reparametrization* of γ if there is an increasing piecewise smooth map of $[c,d]$ onto $[a,b]$ that takes γ to σ, in the sense that $\sigma = \gamma \circ \tau$.

Proposition 4.2. *If $\gamma(t)$, $a \le t \le b$ is a piecewise smooth parametrization of a curve and $\sigma(t)$, $c \le t \le d$ is a reparametrization of γ then*

$$\int_c^d f(\sigma(t))\sigma'(t)\,dt = \int_a^b f(\gamma(t))\gamma'(t)\,dt.$$

Example 4.3. To appreciate this statement, consider the two parametrizations

$$\gamma(t) = e^{it},\ 0 \le t \le 2\pi, \qquad \text{and} \qquad \sigma(t) = e^{2\pi i \sin(t)},\ 0 \le t \le \tfrac{\pi}{2},$$

of the unit circle. Then we could write $\int_\gamma f$ in the two ways

$$\int_\gamma f = i\int_0^{2\pi} f(e^{it})e^{it}\,dt$$

and

$$\int_\gamma f = 2\pi i \int_0^{\frac{\pi}{2}} f\left(e^{2\pi i \sin(t)}\right) e^{2\pi i \sin(t)} \cos(t)\,dt.$$

DEFINITION AND BASIC PROPERTIES

A quick substitution shows that the two integrals on the respective right-hand sides are indeed equal. □

Proposition 4.2 says that a similar equality will hold for any integral and any parametrization. Its proof is left as Exercise 4.9, which also shows that the following definition is unchanged under reparametrization.

Definition. The *length* of a smooth path γ is

$$\text{length}(\gamma) := \int_a^b |\gamma'(t)| \, dt$$

for any parametrization $\gamma(t)$, $a \leq t \leq b$. Naturally, the length of a *piecewise* smooth path is the sum of the lengths of its smooth components.

Example 4.4. Let γ be the line segment from 0 to $1+i$, which can be parametrized by $\gamma(t) = t + it$ for $0 \leq t \leq 1$. Then $\gamma'(t) = 1+i$ and so

$$\text{length}(\gamma) = \int_0^1 |1+i| \, dt = \int_0^1 \sqrt{2} \, dt = \sqrt{2}.$$ □

Example 4.5. Let γ be the unit circle, which can be parametrized by $\gamma(t) = e^{it}$ for $0 \leq t \leq 2\pi$. Then $\gamma'(t) = i e^{it}$ and

$$\text{length}(\gamma) = \int_0^{2\pi} |i e^{it}| \, dt = \int_0^{2\pi} dt = 2\pi.$$ □

Now we observe some basic facts about how the line integral behaves with respect to function addition, scalar multiplication, inverse parametrization, and path concatenation; we also give an upper bound for the absolute value of an integral, which we will make use of time and again.

Proposition 4.6. Suppose γ is a piecewise smooth path, f and g are complex functions which are continuous on γ, and $c \in \mathbb{C}$.

(a) $\int_\gamma (f + c g) = \int_\gamma f + c \int_\gamma g.$

(b) If γ is parametrized by $\gamma(t)$, $a \leq t \leq b$, we define the path $-\gamma$ by $-\gamma(t) := \gamma(a+b-t)$, $a \leq t \leq b$. Then

$$\int_{-\gamma} f = -\int_\gamma f.$$

(c) If γ_1 and γ_2 are piecewise smooth paths so that γ_2 starts where γ_1 ends, we define the path $\gamma_1\gamma_2$ by following γ_1 to its end and then continuing on γ_2 to its end. Then
$$\int_{\gamma_1\gamma_2} f = \int_{\gamma_1} f + \int_{\gamma_2} f.$$

(d) $\left|\int_\gamma f\right| \leq \max_{z \in \gamma} |f(z)| \cdot \text{length}(\gamma).$

The path $-\gamma$ defined in ((b)) is the path that we obtain by traveling through γ in the opposite direction.

Proof. ((a)) follows directly from the definition of the integral and Theorem A.4, the analogous theorem from calculus.

((b)) follows with the real change of variables $s = a + b - t$:
$$\int_{-\gamma} f = \int_a^b f(\gamma(a+b-t))(\gamma(a+b-t))' \, dt$$
$$= -\int_a^b f(\gamma(a+b-t))\gamma'(a+b-t) \, dt$$
$$= \int_b^a f(\gamma(s))\gamma'(s) \, ds = -\int_a^b f(\gamma(s))\gamma'(s) \, ds = -\int_\gamma f.$$

((c)) We need a suitable parametrization $\gamma(t)$ for $\gamma_1\gamma_2$. If γ_1 has domain $[a_1, b_1]$ and γ_2 has domain $[a_2, b_2]$ then we can use
$$\gamma(t) := \begin{cases} \gamma_1(t) & \text{if } a_1 \leq t \leq b_1, \\ \gamma_2(t - b_1 + a_2) & \text{if } b_1 \leq t \leq b_1 + b_2 - a_2, \end{cases}$$

with domain $[a_1, b_1 + b_2 - a_2]$. Now we break the integral over $\gamma_1\gamma_2$ into two pieces and apply the change of variables $s = t - b_1 + a_2$:

$$\int_{\gamma_1\gamma_2} f = \int_{a_1}^{b_1+b_2-a_2} f(\gamma(t))\gamma'(t)\,dt$$

$$= \int_{a_1}^{b_1} f(\gamma(t))\gamma'(t)\,dt + \int_{b_1}^{b_1+b_2-a_2} f(\gamma(t))\gamma'(t)\,dt$$

$$= \int_{\gamma_1} f + \int_{\gamma_2} f.$$

The last step follows since γ restricted to $[a_1, b_1]$ is γ_1 and γ restricted to $[b_1, b_1 + b_2 - a_2]$ is a reparametrization of γ_2.

((d)) Let $\varphi = \left(\operatorname{Arg}\int_\gamma f\right)$. Then $\int_\gamma f = \left|\int_\gamma f\right| e^{i\varphi}$ and thus, since $\left|\int_\gamma f\right| \in \mathbb{R}$,

$$\left|\int_\gamma f\right| = e^{-i\varphi} \int_\gamma f = \operatorname{Re}\left(e^{-i\varphi} \int_a^b f(\gamma(t))\gamma'(t)\,dt\right)$$

$$= \int_a^b \operatorname{Re}\left(f(\gamma(t))e^{-i\varphi}\gamma'(t)\right)dt$$

$$\leq \int_a^b \left|f(\gamma(t))e^{-i\varphi}\gamma'(t)\right|dt = \int_a^b |f(\gamma(t))|\left|\gamma'(t)\right|dt$$

$$\leq \max_{a \leq t \leq b} |f(\gamma(t))| \int_a^b \left|\gamma'(t)\right|dt = \max_{z \in \gamma}|f(z)| \cdot \operatorname{length}(\gamma).$$

Here we have used Theorem A.5 for both inequalities. □

Example 4.7. In Exercise 4.4, you are invited to show

$$\int_\gamma \frac{dz}{z-w} = 2\pi i,$$

where γ is any circle centered at $w \in \mathbb{C}$, oriented counter-clockwise. Thus Proposition 4.6((b)) says that the analogous integral over a *clockwise* circle equals $-2\pi i$. Incidentally, the same example shows that the inequality in Proposition 4.6((d)) is sharp: if γ has radius r, then

$$2\pi = \left|\int_\gamma \frac{dz}{z-w}\right| \leq \max_{z \in \gamma}\left|\frac{1}{z-w}\right| \operatorname{length}(\gamma) = \frac{1}{r} \cdot 2\pi r. \qquad \square$$

4.2 Antiderivatives

The central result about integration of a real function is the Fundamental Theorem of Calculus (Theorem A.3), and our next goal is to discuss complex versions of this theorem. The Fundamental Theorem of Calculus makes a number of important claims: that continuous functions are integrable, their antiderivatives are continuous and differentiable, and that antiderivatives provide easy ways to compute values of definite integrals. The difference between the real case and the complex case is that in the latter, we need to think about integrals over arbitrary paths in $\mathbb{C}$.

Definition. If F is holomorphic in the region $G \subseteq \mathbb{C}$ and $F'(z) = f(z)$ for all $z \in G$, then F is *an antiderivative of f on G*, also known as a *primitive of f on G*.

Example 4.8. We have already seen that $F(z) = z^2$ is entire and has derivative $f(z) = 2z$. Thus, F is an antiderivative of f on any region $G \subseteq \mathbb{C}$. The same goes for $F(z) = z^2 + c$, where $c \in \mathbb{C}$ is any constant. □

Example 4.9. Since

$$\frac{d}{dz}\left(\frac{1}{2i}(\exp(iz) - \exp(-iz))\right) = \frac{1}{2}(\exp(iz) + \exp(-iz)),$$

$F(z) = \sin z$ is an antiderivative of $f(z) = \cos z$ on $\mathbb{C}$. □

Example 4.10. The function $F(z) = \text{Log}(z)$ is an antiderivative of $f(z) = \frac{1}{z}$ on $\mathbb{C} \setminus \mathbb{R}_{\leq 0}$. Note that f is holomorphic in the larger region $\mathbb{C} \setminus \{0\}$; however, we will see in Example 4.14 that f *cannot* have an antiderivative on that region. □

Here is the complex analogue of Theorem A.3((b)).

Theorem 4.11. *Suppose $G \subseteq \mathbb{C}$ is a region and $\gamma \subset G$ is a piecewise smooth path with parametrization $\gamma(t)$, $a \leq t \leq b$. If f is continuous on G and F is any antiderivative of f on G then*

$$\int_\gamma f = F(\gamma(b)) - F(\gamma(a)).$$

Proof. This follows immediately from the definition of a complex integral and Theorem A.3((b)), since $\frac{d}{dt} F(\gamma(t)) = f(\gamma(t)) \gamma'(t)$:

$$\int_\gamma f = \int_a^b f(\gamma(t)) \gamma'(t) \, dt = F(\gamma(b)) - F(\gamma(a)). \qquad \square$$

Example 4.12. Since $F(z) = \frac{1}{2} z^2$ is an antiderivative of $f(z) = z$ in $\mathbb{C}$,

$$\int_\gamma f = \frac{1}{2}(1+i)^2 - \frac{1}{2} 0^2 = i$$

for each of the three paths in Example 4.1. □

There are several interesting consequences of Theorem 4.11. For starters, if γ is closed (that is, $\gamma(a) = \gamma(b)$) we effortlessly obtain the following.

Corollary 4.13. Suppose $G \subseteq \mathbb{C}$ is open, $\gamma \subset G$ is a piecewise smooth closed path, and f is continuous on G and has an antiderivative on G. Then

$$\int_\gamma f = 0.$$

This corollary is immediately useful as a test for existence of antiderivatives:

Example 4.14. The function $f : \mathbb{C} \setminus \{0\} \to \mathbb{C}$ given by $f(z) = \frac{1}{z}$ satisfies $\int_\gamma f = 2\pi i$ for the unit circle $\gamma \subset \mathbb{C} \setminus \{0\}$, by Exercise 4.4. Since this integral is nonzero, f cannot have an antiderivative in $\mathbb{C} \setminus \{0\}$. □

We now turn to the complex analogue of Theorem A.3((a)).

Theorem 4.15. Suppose $G \subseteq \mathbb{C}$ is a region and $z_0 \in G$. Let $f : G \to \mathbb{C}$ be a continuous function such that $\int_\gamma f = 0$ for any closed piecewise smooth path $\gamma \subset G$. Then the function $F : G \to \mathbb{C}$ defined by

$$F(z) := \int_{\gamma_z} f,$$

where γ_z is any piecewise smooth path in G from z_0 to z, is an antiderivative for f on G.

Proof. There are two statements that we have to prove: first, that our definition of F is sound—that is, the integral defining F does not depend on *which* path we take from z_0 to z—and second, that $F'(z) = f(z)$ for all $z \in G$.

Suppose $G \subseteq \mathbb{C}$ is a region, $z_0 \in G$, and $f : G \to \mathbb{C}$ is a continuous function such that $\int_\gamma f = 0$ for any closed piecewise smooth path $\gamma \subset G$. Then $\int_\sigma f$ evaluates to the same number for any piecewise smooth path $\sigma \subset G$ from z_0 to $z \in G$, because any two such paths σ_1 and σ_2 can be concatenated to a closed path first tracing through σ_1 and then through σ_2 backwards, which by assumption yields a zero integral:

$$\int_{\sigma_1} f - \int_{\sigma_2} f = \int_{\sigma_1 - \sigma_2} f = 0.$$

This means that

$$F(z) := \int_{\gamma_z} f$$

is well defined. By the same argument,

$$F(z+h) - F(z) = \int_{\gamma_{z+h}} f - \int_{\gamma_z} f = \int_{\gamma} f$$

for any path $\gamma \subset G$ from z to $z+h$. The constant function 1 has the antiderivative z on $\mathbb{C}$, and so $\int_\gamma 1 = h$, by Theorem 4.11. Thus

$$\frac{F(z+h) - F(z)}{h} - f(z) = \frac{1}{h}\int_\gamma f(w)\,dw - \frac{f(z)}{h}\int_\gamma dw$$

$$= \frac{1}{h}\int_\gamma (f(w) - f(z))\,dw.$$

If $|h|$ is sufficiently small then the line segment λ from z to $z+h$ will be contained in G, and so, by applying the assumptions of our theorem for the third time,

$$\frac{F(z+h) - F(z)}{h} - f(z) = \frac{1}{h}\int_\gamma (f(w) - f(z))\,dw = \frac{1}{h}\int_\lambda (f(w) - f(z))\,dw. \quad (4.2)$$

We will show that the right-hand side goes to zero as $h \to 0$, which will conclude the theorem. Given $\varepsilon > 0$, we can choose $\delta > 0$ such that

$$|w - z| < \delta \quad \Longrightarrow \quad |f(w) - f(z)| < \varepsilon$$

because f is continuous at z. (We also choose δ small enough so that (4.2) holds.) Thus if $|h| < \delta$, we can estimate with Proposition 4.6((d))

$$\left| \frac{1}{h}\int_\lambda (f(w) - f(z))\,dw \right| \leq \frac{1}{|h|} \max_{w \in \lambda} |f(w) - f(z)| \operatorname{length}(\lambda)$$

$$= \max_{w \in \lambda} |f(w) - f(z)| < \varepsilon. \quad \square$$

There are several variations of Theorem 4.15, as we can play with the assumptions about paths in the statement of the theorem. We give one such variation, namely, for *polygonal paths*, i.e., paths that are composed as unions of line segments. You should convince yourself that the proof of the following result is identical to that of Theorem 4.15.

Corollary 4.16. Suppose $G \subseteq \mathbb{C}$ is a region and $z_0 \in G$. Let $f : G \to \mathbb{C}$ be a continuous function such that $\int_\gamma f = 0$ for any closed polygonal path $\gamma \subset G$. Then the function $F : G \to \mathbb{C}$ defined by

$$F(z) := \int_{\gamma_z} f,$$

where γ_z is any polygonal path in G from z_0 to z, is an antiderivative for f on G.

If you compare our proof of Theorem 4.15 to its analogue in $\mathbb{R}$, you will see similarities, as well as some complications due to the fact that we now have to operate in the plane as opposed to the real line. Still, so far we have essentially been "doing calculus" when computing integrals. We will now take a radical departure from this philosophy by studying complex integrals that stay invariant under certain transformations of the paths we are integrating over.

4.3   Cauchy's Theorem

The central theorem of complex analysis is based on the following concept.

Definition. Suppose γ_0 and γ_1 are closed paths in the region $G \subseteq \mathbb{C}$, parametrized by $\gamma_0(t)$, $0 \le t \le 1$, and $\gamma_1(t)$, $0 \le t \le 1$, respectively. Then γ_0 is *G-homotopic* to γ_1 if there exists a continuous function $h : [0,1]^2 \to G$ such that, for all $s, t \in [0,1]$,

$$\begin{aligned} h(t,0) &= \gamma_0(t), \\ h(t,1) &= \gamma_1(t), \\ h(0,s) &= h(1,s). \end{aligned} \quad (4.3)$$

We use the notation $\gamma_1 \sim_G \gamma_2$ to mean γ_1 is G-homotopic to γ_2.

The function $h(t,s)$ is called a *homotopy*. For each fixed s, a homotopy $h(t,s)$ is a path parametrized by t, and as s goes from 0 to 1, these paths continuously

Example 4.17. Figure 4.1 illustrates that the unit circle is $(\mathbb{C} \setminus \{0\})$-homotopic to the square with vertices $\pm 3 \pm 3i$. Indeed, you should check (Exercise 4.20) that

$$h(t,s) := (1-s)e^{2\pi i t} + 3s \times \begin{cases} 1 + 8it & \text{if } 0 \leq t \leq \frac{1}{8}, \\ 2 - 8t + i & \text{if } \frac{1}{8} \leq t \leq \frac{3}{8}, \\ -1 + 4i(1-2t) & \text{if } \frac{3}{8} \leq t \leq \frac{5}{8}, \\ 8t - 6 - i & \text{if } \frac{5}{8} \leq t \leq \frac{7}{8}, \\ 1 + 8i(t-1) & \text{if } \frac{7}{8} \leq t \leq 1 \end{cases} \quad (4.4)$$

gives a homotopy. Note that $h(t,s) \neq 0$ for any $0 \leq t, s \leq 1$ (hence "$(\mathbb{C} \setminus \{0\})$-homotopic"). □

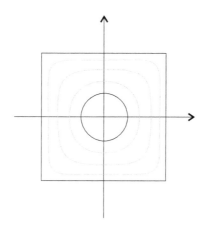

Figure 4.1: This square and circle are $(\mathbb{C} \setminus \{0\})$-homotopic.

Exercise 4.22 shows that $\sim_G$ is an equivalence relation on the set of closed paths in G. The definition of homotopy applies to *parametrizations* of curves; but Exercise 4.23, together with transitivity of $\sim_G$, shows that homotopy is invariant under reparametrizations.

Theorem 4.18 (Cauchy's Theorem). Suppose $G \subseteq \mathbb{C}$ is a region, f is holomorphic in G, γ_0 and γ_1 are piecewise smooth paths in G, and $\gamma_0 \sim_G \gamma_1$. Then

$$\int_{\gamma_0} f = \int_{\gamma_1} f.$$

As a historical aside, it is assumed that Johann Carl Friedrich Gauß (1777–1855) knew a version of this theorem in 1811 but published it only in 1831. Cauchy (of Cauchy–Riemann equations fame) published his version in 1825, Karl Theodor Wilhelm Weierstraß (1815–1897) his in 1842. Theorem 4.18 is often called the *Cauchy–Goursat Theorem*, since Cauchy assumed that the derivative of f was continuous, a condition that was first removed by Edouard Jean-Baptiste Goursat (1858–1936).

Before discussing the proof of Theorem 4.18, we give a basic, yet prototypical application of it:

Example 4.19. We claim that

$$\int_\gamma \frac{dz}{z} = 2\pi i \qquad (4.5)$$

where γ is the square in Figure 4.1, oriented counter-clockwise. We could, of course, compute this integral by hand, but it is easier to apply Cauchy's Theorem 4.18 to the function $f(z) = \frac{1}{z}$, which is holomorphic in $G = \mathbb{C}\setminus\{0\}$. We showed in Example 4.4 that γ is G-homotopic to the unit circle. Exercise 4.4 says that integrating f over the unit circle gives $2\pi i$ and so Cauchy's Theorem 4.18 implies (4.5). □

Proof of Theorem 4.18. The full proof of Cauchy's Theorem is beyond the scope of this book. However, there are several proofs under more restrictive hypotheses than Theorem 4.18. We shall present a proof under the following extra assumptions:

- The derivative f' is continuous in G.

- The homotopy h from γ_0 to γ_1 has piecewise, continuous second derivatives.

Technically, this is the assumption on h:

$$h(t,s) = \begin{cases} h_1(t,s) & \text{if } 0 \leq t \leq t_1, \\ h_2(t,s) & \text{if } t_1 \leq t \leq t_2, \\ \vdots \\ h_n(t,s) & \text{if } t_{n-1} \leq t \leq 1, \end{cases}$$

where each $h_j(t,s)$ has continuous second partials[2]. (Example 4.17 gives one instance.) Now we turn to the proof under these extra assumptions.

For $0 \leq s \leq 1$, let γ_s be the path parametrized by $h(t,s)$, $0 \leq t \leq 1$. Consider the function $I : [0,1] \to \mathbb{C}$ given by

$$I(s) := \int_{\gamma_s} f,$$

so that $I(0) = \int_{\gamma_0} f$ and $I(1) = \int_{\gamma_1} f$. We will show that I is constant; in particular, $I(0) = I(1)$, which proves the theorem. By Leibniz's rule (Theorem A.9),

$$\begin{aligned} \frac{d}{ds} I(s) &= \frac{d}{ds} \int_0^1 f(h(t,s)) \frac{\partial h}{\partial t} \, dt = \int_0^1 \frac{\partial}{\partial s} \left(f(h(t,s)) \frac{\partial h}{\partial t} \right) dt \\ &= \int_0^1 \left(f'(h(t,s)) \frac{\partial h}{\partial s} \frac{\partial h}{\partial t} + f(h(t,s)) \frac{\partial^2 h}{\partial s \, \partial t} \right) dt \\ &= \int_0^1 \left(f'(h(t,s)) \frac{\partial h}{\partial t} \frac{\partial h}{\partial s} + f(h(t,s)) \frac{\partial^2 h}{\partial t \, \partial s} \right) dt \\ &= \int_0^1 \frac{\partial}{\partial t} \left(f(h(t,s)) \frac{\partial h}{\partial s} \right) dt. \end{aligned}$$

Note that we used Theorem A.7 to switch the order of the second partials in the penultimate step—here is where we need our assumption that h has continuous second partials. Also, we needed continuity of f' in order to apply Leibniz's rule. If h is piecewise defined, we split up the integral accordingly.

---

[2] As we have seen with other "piecewise" definitions, the behavior of h at the subdivision lines $t = t_i$ needs to be understood in terms of limits.

Finally, by the Fundamental Theorem of Calculus (Theorem A.3), applied separately to the real and imaginary parts of the above integrand,

$$\frac{d}{ds}I(s) = \int_0^1 \frac{\partial}{\partial t}\left(f(h(t,s))\frac{\partial h}{\partial s}\right)dt$$

$$= f(h(1,s))\frac{\partial h}{\partial s}(1,s) - f(h(0,s))\frac{\partial h}{\partial s}(0,s) = 0,$$

where the last step follows from $h(0,s) = h(1,s)$ for all s. □

Definition. Let $G \subseteq \mathbb{C}$ be a region. If the closed path γ is G-homotopic to a point (that is, a constant path) then γ is *G-contractible*, and we write $\gamma \sim_G 0$. (See Figure 4.2 for an example.)

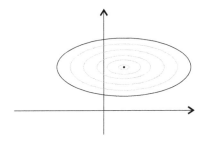

Figure 4.2: This ellipse is $(\mathbb{C} \setminus \mathbb{R})$-contractible.

The fact that an integral over a point is zero has the following immediate consequence.

Corollary 4.20. Suppose $G \subseteq \mathbb{C}$ is a region, f is holomorphic in G, γ is piecewise smooth, and $\gamma \sim_G 0$. Then

$$\int_\gamma f = 0.$$

This corollary is worth meditating over. For example, you should compare it with Corollary 4.13: both results give a zero integral, yet they make truly opposite assumptions (one about the existence of an antiderivative, the other about the existence of a derivative).

Naturally, Corollary 4.20 gives many evaluations of integrals, such as this:

Example 4.21. Since Log is holomorphic in $G = \mathbb{C} \setminus \mathbb{R}_{\leq 0}$ and the ellipse γ in Figure 4.2 is G-contractible, Corollary 4.20 gives

$$\int_\gamma \mathrm{Log}(z)\,dz = 0.$$

□

Exercise 4.24(a) says that any closed path is $\mathbb{C}$-contractible, which yields the following special case of Corollary 4.20.

Corollary 4.22. If f is entire and γ is any piecewise smooth closed path, then

$$\int_\gamma f = 0.$$

The theorems and corollaries in this section are useful not just for showing that certain integrals are zero:

Example 4.23. We'd like to compute

$$\int_\gamma \frac{dz}{z^2 - 2z}$$

where γ is the unit circle, oriented counter-clockwise. (Try computing it from first principles.) We use a partial fractions expansion to write

$$\int_\gamma \frac{dz}{z^2 - 2z} = \frac{1}{2} \int_\gamma \frac{dz}{z - 2} - \frac{1}{2} \int_\gamma \frac{dz}{z}.$$

The first integral on the right-hand side is zero by Corollary 4.20 applied to the function $f(z) = \frac{1}{z-2}$ (note that f is holomorphic in $\mathbb{C} \setminus \{2\}$ and γ is $(\mathbb{C} \setminus \{2\})$-contractible). The second integral is $2\pi i$ by Exercise 4.4, and so

$$\int_\gamma \frac{dz}{z^2 - 2z} = -\pi i.$$

□

Sometimes Corollary 4.20 itself is known as Cauchy's Theorem. See Exercise 4.25 for a related formulation of Corollary 4.20, with a proof based on Green's Theorem.

4.4 Cauchy's Integral Formula

We recall our notations

$$C[a,r] = \{z \in \mathbb{C} : |z-a| = r\}$$
$$D[a,r] = \{z \in \mathbb{C} : |z-a| < r\}$$
$$\overline{D}[a,r] = \{z \in \mathbb{C} : |z-a| \leq r\}$$

for the circle, open disk, and closed disk, respectively, with center $a \in \mathbb{C}$ and radius $r > 0$. Unless stated otherwise, we orient $C[a,r]$ counter-clockwise.

Theorem 4.24. *If f is holomorphic in an open set containing $\overline{D}[w,R]$ then*

$$f(w) = \frac{1}{2\pi i} \int_{C[w,R]} \frac{f(z)}{z-w}\, dz.$$

This is *Cauchy's Integral Formula* for the case that the integration path is a circle; we will prove the general statement at the end of this chapter. However, already this special case is worth meditating over: the data on the right-hand side of Theorem 4.24 is entirely given by the values that $f(z)$ takes on for z on the circle $C[w,R]$. Thus Cauchy's Integral Formula says that this data determines $f(w)$. This has the flavor of *mean-value theorems*, which the following corollary makes even more apparent.

Corollary 4.25. *If $f = u + iv$ is holomorphic in an open set containing $\overline{D}[w,R]$ then*

$$f(w) = \frac{1}{2\pi} \int_0^{2\pi} f\!\left(w + R\,e^{it}\right) dt,$$

$$u(w) = \frac{1}{2\pi} \int_0^{2\pi} u\!\left(w + R\,e^{it}\right) dt \quad \text{and} \quad v(w) = \frac{1}{2\pi} \int_0^{2\pi} v\!\left(w + R\,e^{it}\right) dt.$$

Proofs of Theorem 4.24 and Corollary 4.25. By assumption, f is holomorphic in an open set G that contains $\overline{D}[w,R]$, and so $\frac{f(z)}{z-w}$ is holomorphic in $H := G \setminus \{w\}$. For any $0 < r < R$,

$$C[w,r] \sim_H C[w,R],$$

and so Cauchy's Theorem 4.18 and Exercise 4.4 give

$$\left| \int_{C[w,R]} \frac{f(z)}{z-w} \, dz - 2\pi i \, f(w) \right| = \left| \int_{C[w,r]} \frac{f(z)}{z-w} \, dz - f(w) \int_{C[w,r]} \frac{dz}{z-w} \right|$$

$$= \left| \int_{C[w,r]} \frac{f(z)-f(w)}{z-w} \, dz \right|$$

$$\leq \max_{z \in C[w,r]} \left| \frac{f(z)-f(w)}{z-w} \right| \operatorname{length}(C[w,r]) \quad (4.6)$$

$$= \max_{z \in C[w,r]} \frac{|f(z)-f(w)|}{r} \, 2\pi r$$

$$= 2\pi \max_{z \in C[w,r]} |f(z)-f(w)|.$$

Here the inequality comes from Proposition 4.6((d)).

Now let $\varepsilon > 0$. Because f is continuous at w, there exists $\delta > 0$ such that $|z - w| < \delta$ implies

$$|f(z) - f(w)| < \frac{\varepsilon}{2\pi}.$$

In particular, this will hold for $z \in C[w, \frac{\delta}{2}]$, and so (4.6) implies, with $r = \frac{\delta}{2}$,

$$\left| \int_{C[w,R]} \frac{f(z)}{z-w} \, dz - 2\pi i \, f(w) \right| < \varepsilon.$$

Since we can choose ε as small as we'd like, the left-hand side must be zero, which proves Theorem 4.24.

Corollary 4.25 now follows by definition of the complex integral:

$$f(w) = \frac{1}{2\pi i} \int_0^{2\pi} \frac{f(w + R e^{it})}{w + R e^{it} - w} \, iR e^{it} \, dt = \frac{1}{2\pi} \int_0^{2\pi} f(w + R e^{it}) \, dt,$$

which splits into real and imaginary parts as

$$u(w) + i \, v(w) = \frac{1}{2\pi} \int_0^{2\pi} u(w + R e^{it}) \, dt + i \frac{1}{2\pi} \int_0^{2\pi} v(w + R e^{it}) \, dt. \quad \square$$

Theorem 4.24 can be used to compute integrals of a certain nature.

Example 4.26. We'd like to determine

$$\int_{C[i,1]} \frac{dz}{z^2 + 1}.$$

The function $f(z) = \frac{1}{z+i}$ is holomorphic in $\mathbb{C} \setminus \{-i\}$, which contains $\overline{D}[i, 1]$. Thus we can apply Theorem 4.24:

$$\int_{C[i,1]} \frac{dz}{z^2 + 1} = \int_{C[i,1]} \frac{\frac{1}{z+i}}{z - i} dz = 2\pi i \, f(i) = 2\pi i \frac{1}{2i} = \pi. \qquad \square$$

Now we would like to extend Theorem 4.24 by replacing $C[w, R]$ with any simple closed piecewise smooth path γ around w. Intuitively, Cauchy's Theorem 4.18 should supply such an extension: assuming that f is holomorphic in a region G that includes γ and its inside, we can find a small R such that $\overline{D}[w, R] \subseteq G$, and since $\frac{f(z)}{z-w}$ is holomorphic in $H := G \setminus \{w\}$ and $\gamma \sim_H C[w, R]$, Theorems 4.18 and 4.24 yield

$$f(w) = \frac{1}{2\pi i} \int_\gamma \frac{f(z)}{z - w} dz.$$

This all smells like good coffee, except ... we might be just dreaming. The argument may be intuitively clear, but intuition doesn't prove anything. We'll look at it carefully, fill in the gaps, and then we'll see what we have proved.

First, we need a notion of the *inside* of a simple closed path. The fact that any such path γ divides the complex plane into two connected open sets of γ (the bounded one of which we call the *inside* or **interior** of γ) is one of the first substantial theorems ever proved in topology, the *Jordan Curve Theorem*, due to Camille Jordan (1838–1922).[3] In this book we shall assume the validity of the Jordan Curve Theorem.

Second, we need to specify the orientation of γ, since if the formula gives $f(w)$ for one orientation then it will give $-f(w)$ for the other orientation.

Definition. A piecewise smooth simple closed path γ is *positively oriented* if it is parametrized so that its inside is on the left as our parametrization traverses γ. An example is a counter-clockwise oriented circle.

Third, if γ is positively oriented and $\overline{D}[w, R]$ is a closed disk inside γ then we need a homotopy from γ to the counterclockwise circle $C[w, R]$ that stays inside γ and away from $D[w, R]$. This is provided directly by another substantial theorem of topology, the *Annulus Theorem*, although there are other methods. Again, in this book we shall assume the existence of this homotopy.

---

[3] This is the Jordan of *Jordan normal form* fame, but not the one of *Gauß–Jordan elimination*.

These results of topology seem intuitively obvious but are surprisingly difficult to prove. If you'd like to see a proof, we recommend that you take a course in topology.

There is still a subtle problem with our proof. We assumed that γ is in G, but we also need the *interior* of γ to be contained in G, since we need to apply Cauchy's Theorem to the homotopy between γ and $C[w,R]$. We could just add this as an assumption to our theorem, but the following formulation will be more convenient later.

Theorem 4.27 (Cauchy's Integral Formula). Suppose f is holomorphic in the region G and γ is a positively oriented, simple, closed, piecewise smooth path, such that w is inside γ and $\gamma \sim_G 0$. Then

$$f(w) = \frac{1}{2\pi i} \int_\gamma \frac{f(z)}{z-w}\,dz.$$

So all that we need to finish the proof of Theorem 4.27 is one more fact from topology. But we can prove this one:

Proposition 4.28. Suppose γ is a simple, closed, piecewise smooth path in the region G. Then G contains the interior of γ if and only if $\gamma \sim_G 0$.

Proof. One direction is easy: If G contains the interior of γ and $\overline{D}[w,R]$ is any closed disk in the interior of γ then there is a G-homotopy from γ to $C[w,R]$, and $C[w,R] \sim_G 0$.

In the other direction we argue by contradiction: Assume $\gamma \sim_G 0$ but G does not contain the interior of γ. So we can find a point w in the interior of γ which is not in G.

Define $g(z) = \frac{1}{z-w}$ for $z \neq w$. Now g is holomorphic on G and $\gamma \sim_G 0$, so Corollary 4.20 applies, and we have $\int_\gamma g(z)\,dz = 0$. On the other hand, choose $R > 0$ so that $\overline{D}[w,R]$ is inside γ. There is a homotopy in $\mathbb{C}\setminus\{w\}$ from γ to $C[w,R]$, so Cauchy's Theorem 4.18, plus Exercise 4.4, shows that $\int_\gamma g(z)\,dz = 2\pi i$.

This contradiction finishes the proof. □

Notice that, instead of using topology to prove a theorem about holomorphic functions, we just used holomorphic functions to prove a theorem about topology.

Example 4.29. Continuing Example 4.26, Theorem 4.27 says that

$$\int_\gamma \frac{dz}{z^2+1} = \pi$$

for any positively oriented, simple, closed, piecewise smooth path γ that contains i on its inside and that is $(\mathbb{C} \setminus \{-i\})$-contractible. □

Example 4.30. To compute

$$\int_{C[0,3]} \frac{\exp(z)}{z^2 - 2z} \, dz$$

we use the partial fractions expansion from Example 4.23:

$$\int_{C[0,3]} \frac{\exp(z)}{z^2 - 2z} \, dz = \frac{1}{2} \int_{C[0,3]} \frac{\exp(z)}{z - 2} \, dz - \frac{1}{2} \int_{C[0,3]} \frac{\exp(z)}{z} \, dz \, .$$

For the two integrals on the right-hand side, we can use Theorem 4.24 with the function $f(z) = \exp(z)$, which is entire, and so (note that both 2 and 0 are inside γ)

$$\int_{C[0,3]} \frac{\exp(z)}{z^2 - 2z} \, dz = \frac{1}{2} 2\pi i \cdot \exp(2) - \frac{1}{2} 2\pi i \cdot \exp(0) = \pi i \left(e^2 - 1 \right). \quad □$$

Exercises

4.1. Find the length of the following paths:

(a) $\gamma(t) = 3t + i, \; -1 \leq t \leq 1$

(b) $\gamma(t) = i + e^{i\pi t}, \; 0 \leq t \leq 1$

(c) $\gamma(t) = i \sin(t), \; -\pi \leq t \leq \pi$

(d) $\gamma(t) = t - i e^{-it}, \; 0 \leq t \leq 2\pi$

Draw pictures of each path and convince yourself that the lengths you computed are sensible. (The last path is a *cycloid*, the trace of a fixed point on a wheel as it makes one rotation.)

4.2. Compute the lengths of the paths from Exercise 1.33:

(a) the circle $C[1+i, 1]$

(b) the line segment from $-1 - i$ to $2i$

(c) the top half of the circle $C[0, 34]$

(d) the rectangle with vertices $\pm 1 \pm 2i$

4.3. Integrate the function $f(z) = \bar{z}$ over the three paths given in Example 4.1.

4.4. Compute $\int_\gamma \frac{dz}{z}$ where γ is the unit circle, oriented counterclockwise. More generally, show that for any $w \in \mathbb{C}$ and $r > 0$,

$$\int_{C[w,r]} \frac{dz}{z-w} = 2\pi i.$$

4.5. Integrate the following functions over the circle $C[0,2]$:

(a) $f(z) = z + \bar{z}$ (c) $f(z) = \frac{1}{z^4}$

(b) $f(z) = z^2 - 2z + 3$ (d) $f(z) = xy$

4.6. Evaluate the integrals $\int_\gamma x\, dz$, $\int_\gamma y\, dz$, $\int_\gamma z\, dz$ and $\int_\gamma \bar{z}\, dz$ along each of the following paths. (*Hint:* You can get the second two integrals after you calculate the first two by writing z and $\bar{z}$ as $x \pm iy$.)

(a) γ is the line segment from 0 to $1-i$

(b) $\gamma = C[0,1]$

(c) $\gamma = C[a, r]$ for some $a \in \mathbb{C}$

4.7. Evaluate $\int_\gamma \exp(3z)\, dz$ for each of the following paths:

(a) γ is the line segment from 1 to i

(b) $\gamma = C[0,3]$

(c) γ is the arc of the parabola $y = x^2$ from $x = 0$ to $x = 1$

4.8. Compute $\int_\gamma f$ for the following functions f and paths γ:

(a) $f(z) = z^2$ and $\gamma(t) = t + it^2$, $0 \le t \le 1$.

(b) $f(z) = z$ and γ is the semicircle from 1 through i to -1.

(c) $f(z) = \exp(z)$ and γ is the line segment from 0 to a point z_0.

(d) $f(z) = |z|^2$ and γ is the line segment from 2 to $3 + i$.

(e) $f(z) = z + \frac{1}{z}$ and γ is parametrized by $\gamma(t)$, $0 \leq t \leq 1$, and satisfies $\operatorname{Im} \gamma(t) > 0$, $\gamma(0) = -4 + i$, and $\gamma(1) = 6 + 2i$.

(f) $f(z) = \sin(z)$ and γ is some piecewise smooth path from i to π.

4.9. Prove Proposition 4.2 and the fact that the length of γ does not change under reparametrization. (*Hint*: Assume γ, σ, and τ are smooth. Start with the definition of $\int_\sigma f$, apply the chain rule to $\sigma = \gamma \circ \tau$, and then use the change of variables formula, Theorem A.6.)

4.10. Prove the following *integration by parts* statement: Let f and g be holomorphic in G, and suppose $\gamma \subset G$ is a piecewise smooth path from $\gamma(a)$ to $\gamma(b)$. Then

$$\int_\gamma f g' = f(\gamma(b))g(\gamma(b)) - f(\gamma(a))g(\gamma(a)) - \int_\gamma f' g.$$

4.11. Let $I(k) := \frac{1}{2\pi} \int_0^{2\pi} e^{ikt}\, dt$.
  (a) Show that $I(0) = 1$.
  (b) Show that $I(k) = 0$ if k is a nonzero integer.
  (c) What is $I(\frac{1}{2})$?

4.12. Compute $\int_{C[0,2]} z^{\frac{1}{2}}\, dz$.

4.13. Show that $\int_\gamma z^n\, dz = 0$ for any closed piecewise smooth γ and any integer $n \neq -1$. (If n is negative, assume that γ does not pass through the origin, since otherwise the integral is not defined.)

4.14. Exercise 4.13 excluded $n = -1$ for a good reason: Exercise 4.4 gives a counterexample. Generalizing these, if m is any integer, find a closed path γ so that $\int_\gamma z^{-1}\, dz = 2m\pi i$.

4.15. Taking the previous two exercises one step further, fix $z_0 \in \mathbb{C}$ and let γ be a simple, closed, positively oriented, piecewise smooth path such that z_0 is inside γ. Show that, for any integer n,

$$\int_\gamma (z-z_0)^n \, dz = \begin{cases} 2\pi i & \text{if } n=-1, \\ 0 & \text{otherwise.} \end{cases}$$

4.16. Prove that $\int_\gamma z \exp(z^2) \, dz = 0$ for any closed path γ.

4.17. Show that $F(z) = \frac{i}{2}\operatorname{Log}(z+i) - \frac{i}{2}\operatorname{Log}(z-i)$ is an antiderivative of $\frac{1}{1+z^2}$ for $\operatorname{Re}(z) > 0$. Is $F(z)$ equal to $\arctan z$?

4.18. Compute the following integrals, where γ is the line segment from 4 to $4i$.

(a) $\displaystyle\int_\gamma \frac{z+1}{z} \, dz$ \qquad (c) $\displaystyle\int_\gamma z^{-\frac{1}{2}} \, dz$

(b) $\displaystyle\int_\gamma \frac{dz}{z^2+z}$ \qquad (d) $\displaystyle\int_\gamma \sin^2(z) \, dz$

4.19. Compute the following integrals. (*Hint*: One of these integrals is considerably easier than the other.)

(a) $\displaystyle\int_{\gamma_1} z^i \, dz$ where $\gamma_1(t) = e^{it}$, $-\frac{\pi}{2} \leq t \leq \frac{\pi}{2}$.

(b) $\displaystyle\int_{\gamma_2} z^i \, dz$ where $\gamma_2(t) = e^{it}$, $\frac{\pi}{2} \leq t \leq \frac{3\pi}{2}$.

4.20. Show that (4.4) gives a homotopy between the unit circle and the square with vertices $\pm 3 \pm 3i$.

4.21. Suppose $a \in \mathbb{C}$ and γ_0 and γ_1 are two counterclockwise circles so that a is inside both of them. Give a homotopy that proves $\gamma_0 \sim_{\mathbb{C}\setminus\{a\}} \gamma_1$.

4.22. Prove that $\sim_G$ is an equivalence relation.

4.23. Suppose that γ is a closed path in a region G, parametrized by $\gamma(t)$, $t \in [0,1]$, and τ is a continuous increasing function from $[0,1]$ onto $[0,1]$. Show that γ is G-homotopic to the reparametrized path $\gamma \circ \tau$. (*Hint*: Make use of $\tau_s(t) = s\tau(t) + (1-s)t$ for $0 \le s \le 1$.)

4.24.

(a) Prove that any closed path is $\mathbb{C}$-contractible.

(b) Prove that any two closed paths are $\mathbb{C}$-homotopic.

4.25. This exercise gives an alternative proof of Corollary 4.20 via Green's Theorem A.10. Suppose $G \subseteq \mathbb{C}$ is a region, f is holomorphic in G, f' is continuous, γ is a simple piecewise smooth closed curve, and $\gamma \sim_G 0$. Explain that we may write

$$\int_\gamma f(z)\,dz = \int_\gamma (u+iv)(dx+i\,dy) = \int_\gamma u\,dx - v\,dy + i\int_\gamma v\,dx + u\,dy$$

and show that these integrals vanish, by using Green's Theorem A.10 together with Proposition 4.28, and then the Cauchy–Riemann equations (2.2).

4.26. Fix $a \in \mathbb{C}$. Compute

$$I(r) := \int_{C[0,r]} \frac{dz}{z-a}.$$

You should get different answers for $r < |a|$ and $r > |a|$. (*Hint*: In one case γ_r is contractible in $\mathbb{C} \setminus \{a\}$. In the other you can combine Exercises 4.4 and 4.21.)

4.27. Suppose $p(z)$ is a polynomial in z and γ is a closed piecewise smooth path in $\mathbb{C}$. Show that $\int_\gamma p = 0$.

4.28. Show that

$$\int_{C[0,2]} \frac{dz}{z^3+1} = 0$$

by arguing that this integral does not change if we replace $C[0,2]$ by $C[0,r]$ for any $r > 1$, then use Proposition 4.6((d)) to obtain an upper bound for $\left|\int_{C[0,r]} \frac{dz}{z^3+1}\right|$ that goes to 0 as $r \to \infty$.

4.29. Compute the *real* integral

$$\int_0^{2\pi} \frac{d\varphi}{2+\sin\varphi}$$

by writing the sine function in terms of the exponential function and making the substitution $z = e^{i\varphi}$ to turn the real integral into a complex integral.

4.30. Prove that for $0 < r < 1$,

$$\frac{1}{2\pi}\int_0^{2\pi} \frac{1-r^2}{1-2r\cos(\varphi)+r^2}\, d\varphi = 1.$$

(The function $P_r(\varphi) := \frac{1-r^2}{1-2r\cos(\varphi)+r^2}$ is the *Poisson kernel*[4] and plays an important role in the world of harmonic functions, as we will see in Exercise 6.13.)

4.31. Suppose f and g are holomorphic in the region G and γ is a simple piecewise smooth G-contractible path. Prove that if $f(z) = g(z)$ for all $z \in \gamma$, then $f(z) = g(z)$ for all z inside γ.

4.32. Show that Corollary 4.20, for simple paths, is also a corollary of Theorem 4.27.

4.33. Compute

$$I(r) := \int_{C[-2i,r]} \frac{dz}{z^2+1}$$

for $r \neq 1, 3$.

4.34. Find

$$\int_{C[0,r]} \frac{dz}{z^2-2z-8}$$

for $r = 1$, $r = 3$ and $r = 5$. (*Hint*: Compute a partial-fractions expansion of the integrand.)

4.35. Use the Cauchy Integral Formula (Theorem 4.24) to evaluate the integral in Exercise 4.34 when $r = 3$.

---

[4]Named after Siméon Denis Poisson (1781–1840).

4.36. Compute the following integrals.

(a) $\int_{C[-1,2]} \dfrac{z^2}{4-z^2}\, dz$

(b) $\int_{C[0,1]} \dfrac{\sin z}{z}\, dz$

(c) $\int_{C[0,2]} \dfrac{\exp(z)}{z(z-3)}\, dz$

(d) $\int_{C[0,4]} \dfrac{\exp(z)}{z(z-3)}\, dz$

4.37. Let $f(z) = \dfrac{1}{z^2-1}$ and define the two paths $\gamma = C[1,1]$ oriented counterclockwise and $\sigma = C[-1,1]$ oriented clockwise. Show that $\int_\gamma f = \int_\sigma f$ even though $\gamma \not\sim_G \sigma$ where $G = \mathbb{C}\setminus\{\pm 1\}$, the region of holomorphicity of f.

4.38. This exercise gives an alternative proof of Cauchy's Integral Formula (Theorem 4.27) that does not depend on Cauchy's Theorem (Theorem 4.18). Suppose the region G is *convex*; this means that, whenever z and w are in G, the line segment between them is also in G. Suppose f is holomorphic in G, f' is continuous, and γ is a positively oriented, simple, closed, piecewise smooth path, such that w is inside γ and $\gamma \sim_G 0$.

(a) Consider the function $g : [0,1] \to \mathbb{C}$ given by

$$g(t) := \int_\gamma \dfrac{f(w + t(z-w))}{z-w}\, dz\,.$$

Show that $g' = 0$. (*Hint*: Use Theorem A.9 (Leibniz's rule) and then find an antiderivative for $\dfrac{\partial f}{\partial t}(z + t(w-z))$.)

(b) Prove Theorem 4.27 by evaluating $g(0)$ and $g(1)$.

(c) Why did we assume G is convex?

Chapter 5

Consequences of Cauchy's Theorem

> *Everybody knows that mathematics is about miracles, only mathematicians have a name for them: theorems.*
> Roger Howe

Cauchy's Theorem and Integral Formula (Theorems 4.18 and 4.27), which we now have at our fingertips, are not just beautiful results but also incredibly practical. In a quite concrete sense, the rest of this book will reap the fruits that these two theorems provide us with. This chapter starts with a few highlights.

5.1 Variations of a Theme

We now derive formulas for f' and f'' which resemble Cauchy's Integral Formula (Theorem 4.27).

Theorem 5.1. Suppose f is holomorphic in the region G and γ is a positively oriented, simple, closed, piecewise smooth, G-contractible path. If w is inside γ then
$$f'(w) = \frac{1}{2\pi i} \int_\gamma \frac{f(z)}{(z-w)^2}\, dz.$$
Moreover, $f''(w)$ exists, and
$$f''(w) = \frac{1}{\pi i} \int_\gamma \frac{f(z)}{(z-w)^3}\, dz.$$

Proof. The idea of our proof is very similar to that of Cauchy's Integral Formula (Theorems 4.24 and 4.27). We will study the following difference quotient, which we rewrite using Theorem 4.27.

$$\begin{aligned}\frac{f(w+\Delta w)-f(w)}{\Delta w} &= \frac{1}{\Delta w}\left(\frac{1}{2\pi i}\int_\gamma \frac{f(z)}{z-(w+\Delta w)}dz - \frac{1}{2\pi i}\int_\gamma \frac{f(z)}{z-w}dz\right) \\ &= \frac{1}{2\pi i}\int_\gamma \frac{f(z)}{(z-w-\Delta w)(z-w)}dz.\end{aligned}$$

Theorem 5.1 will follow if we can show that the following expression gets arbitrarily small as $\Delta w \to 0$:

$$\begin{aligned}&\frac{f(w+\Delta w)-f(w)}{\Delta w} - \frac{1}{2\pi i}\int_\gamma \frac{f(z)}{(z-w)^2}dz \\ &= \frac{1}{2\pi i}\int_\gamma \left(\frac{f(z)}{(z-w-\Delta w)(z-w)} - \frac{f(z)}{(z-w)^2}\right)dz \\ &= \frac{\Delta w}{2\pi i}\int_\gamma \frac{f(z)}{(z-w-\Delta w)(z-w)^2}dz. \quad (5.1)\end{aligned}$$

This can be made arbitrarily small if we can show that the integral on the right-hand side stays bounded as $\Delta w \to 0$. In fact, by Proposition 4.6((d)), it suffices to show that the *integrand* stays bounded as $\Delta w \to 0$ (because γ and hence length(γ) are fixed).

Let $M := \max_{z\in\gamma}|f(z)|$ (whose existence is guaranteed by Theorem A.1). Choose $\delta > 0$ such that $D[w,\delta]\cap \gamma = \emptyset$; that is, $|z-w| \geq \delta$ for all z on γ. By the reverse triangle inequality (Corollary 1.7((b))), for all $z \in \gamma$,

$$\left|\frac{f(z)}{(z-w-\Delta w)(z-w)^2}\right| \leq \frac{|f(z)|}{(|z-w|-|\Delta w|)|z-w|^2} \leq \frac{M}{(\delta-|\Delta w|)\delta^2},$$

which certainly stays bounded as $\Delta w \to 0$. This proves (5.1) and thus the Cauchy Integral Formula for f'.

The proof of the formula for f'' is very similar and will be left to Exercise 5.2. □

Theorem 5.1 suggests that there are similar formulas for the higher derivatives of f. This is in fact true, and theoretically we could obtain them one by one with the methods of the proof of Theorem 5.1. However, once we start studying power series for holomorphic functions, we will obtain such a result much more easily; so we save the derivation of integral formulas for higher derivatives of f for later (Corollary 8.11).

Theorem 5.1 has several important consequences. For starters, it can be used to compute certain integrals.

Example 5.2.

$$\int_{C[0,1]} \frac{\sin(z)}{z^2} \, dz = 2\pi i \left. \frac{d}{dz} \sin(z) \right|_{z=0} = 2\pi i \cos(0) = 2\pi i. \qquad \square$$

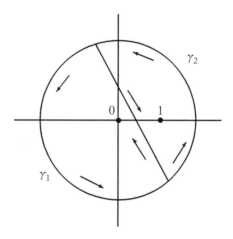

Figure 5.1: The integration paths in Example 5.3.

Example 5.3. To compute the integral

$$\int_{C[0,2]} \frac{dz}{z^2(z-1)},$$

we could employ a partial fractions expansion similar to the one in Example 4.23, or moving the integration path similar to the one in Exercise 4.28. To exhibit an alternative, we split up the integration path as illustrated in Figure 5.1: we introduce an additional path that separates 0 and 1. If we integrate on these two new closed paths (γ_1 and γ_2) counterclockwise, the two contributions along the new path will cancel each other. The effect is that we transformed an integral for which two singularities were inside the integration path into a sum of two integrals, each of which has only one singularity inside the integration path; these new integrals we know how to deal with, using Theorems 4.24 and 5.1:

$$\int_{C[0,2]} \frac{dz}{z^2(z-1)} = \int_{\gamma_1} \frac{dz}{z^2(z-1)} + \int_{\gamma_2} \frac{dz}{z^2(z-1)} = \int_{\gamma_1} \frac{\frac{1}{z-1}}{z^2} dz + \int_{\gamma_2} \frac{\frac{1}{z^2}}{z-1} dz$$

$$= 2\pi i \left. \frac{d}{dz} \frac{1}{z-1} \right|_{z=0} + 2\pi i \frac{1}{1^2} = 2\pi i \left(-\frac{1}{(-1)^2} \right) + 2\pi i$$

$$= 0. \qquad \square$$

Example 5.4.

$$\int_{C[0,1]} \frac{\cos(z)}{z^3} dz = \pi i \left. \frac{d^2}{dz^2} \cos(z) \right|_{z=0} = \pi i (-\cos(0)) = -\pi i. \qquad \square$$

Theorem 5.1 has another powerful consequence: just from knowing that f is holomorphic in G, we know of the existence of f'', that is, f' is also holomorphic in G. Repeating this argument for f', then for f'', f''', etc., shows that all derivatives $f^{(n)}$ exist and are holomorphic. We can translate this into the language of partial derivatives, since the Cauchy–Riemann equations (Theorem 2.13) show that any sequence of n partial differentiations of f results in a constant times $f^{(n)}$.

So we have the following statement, which has no analogue whatsoever in the reals (see, e.g., Exercise 5.6).

Corollary 5.5. If f is differentiable in a region G then f is infinitely differentiable in G, and all partials of f with respect to x and y exist and are continuous.

5.2 Antiderivatives Again

Theorem 4.15 gave us an antiderivative for a function that has zero integrals over closed paths in a given region. Now that we have Corollary 5.5, meditating just a bit more over Theorem 4.15 gives a converse of sorts to Corollary 4.20.

Corollary 5.6 (Morera's[1] Theorem). Suppose f is continuous in the region G and

$$\int_\gamma f = 0$$

for all piecewise smooth closed paths $\gamma \subset G$. Then f is holomorphic in G.

---
[1] Named after Giancinto Morera (1856–1907).

Proof. Theorem 4.15 yields an antiderivative F for f in G. Because F is holomorphic in G, Corollary 5.5 implies that f is also holomorphic in G. □

Just like there are several variations of Theorem 4.15, we have variations of Corollary 5.6. For example, by Corollary 4.16, we can replace the condition *for all piecewise smooth closed paths* $\gamma \subset G$ in the statement of Corollary 5.6 by the condition *for all closed polygonal paths* $\gamma \subset G$ (which, in fact, gives a stronger version of this result).

A special case of Theorem 4.15 applies to regions in which every closed path is contractible.

Definition. A region $G \subseteq \mathbb{C}$ is *simply connected* if $\gamma \sim_G 0$ for every closed path γ in G.

Loosely speaking, a region is simply connected if it has no holes.

Example 5.7. Any disk $D[a, r]$ is simply connected, as is $\mathbb{C} \setminus \mathbb{R}_{\leq 0}$. (You should draw a few closed paths in $\mathbb{C} \setminus \mathbb{R}_{\leq 0}$ to convince yourself that they are all contractible.) The region $\mathbb{C} \setminus \{0\}$ is not simply connected as, e.g., the unit circle is not $(\mathbb{C} \setminus \{0\})$-contractible. □

If f is holomorphic in a simply-connected region then Corollary 4.20 implies that f satisfies the conditions of Theorem 4.15, whence we conclude:

Corollary 5.8. *Every holomorphic function on a simply-connected region $G \subseteq \mathbb{C}$ has an antiderivative on G.*

Note that this corollary gives no indication of how to compute an antiderivative. For example, it says that the (entire) function $f : \mathbb{C} \to \mathbb{C}$ given by $f(z) = \exp(z^2)$ has an antiderivative F in $\mathbb{C}$; it is an entirely different matter to derive a formula for F.

Corollary 5.8 also illustrates the role played by two of the regions in Example 5.7, in connection with the function $f(z) = \frac{1}{z}$. This function has no antiderivative on $\mathbb{C} \setminus \{0\}$, as we proved in Example 4.14. Consequently (as one can see much more easily), $\mathbb{C} \setminus \{0\}$ is not simply connected. However, the function $f(z) = \frac{1}{z}$ does have an antiderivative on the simply-connected region $\mathbb{C} \setminus \mathbb{R}_{\leq 0}$ (namely, $\text{Log}(z)$), illustrating one instance implied by Corollary 5.8.

Finally, Corollary 5.8 implies that, if we have two paths in a simply-connected region with the same endpoints, we can concatenate them—changing direction on one—to form a closed path, which proves:

Corollary 5.9. If f is holomorphic in a simply-connected region G then $\int_\gamma f$ is independent of the piecewise smooth path $\gamma \subset G$ between $\gamma(a)$ and $\gamma(b)$.

When an integral depends only on the endpoints of the path, the integral is called *path independent*. Example 4.1 shows that this situation is quite special; it also says that the function $\bar{z}^2$ does not have an antiderivative in, for example, the region $\{z \in \mathbb{C} : |z| < 2\}$. (Actually, the function $\bar{z}^2$ does not have an antiderivative in any nonempty region—see Exercise 5.7.)

5.3 Taking Cauchy's Formulas to the Limit

Many beautiful applications of Cauchy's Integral Formulas (such as Theorems 4.27 and 5.1) arise from considerations of the limiting behavior of the integral as the path gets arbitrarily large. The first and most famous application concerns the roots of polynomials. As a preparation we prove the following inequality, which is generally quite useful. It says that for $|z|$ large enough, a polynomial $p(z)$ of degree d looks almost like a constant times z^d.

Proposition 5.10. *Suppose $p(z)$ is a polynomial of degree d with leading coefficient a_d. Then there is a real number R such that*

$$\tfrac{1}{2} |a_d| |z|^d \leq |p(z)| \leq 2 |a_d| |z|^d$$

for all z satisfying $|z| \geq R$.

Proof. Since $p(z)$ has degree d, its leading coefficient a_d is not zero, and we can factor out $a_d z^d$:

$$|p(z)| = \left| a_d z^d + a_{d-1} z^{d-1} + a_{d-2} z^{d-2} + \cdots + a_1 z + a_0 \right|$$
$$= |a_d| |z|^d \left| 1 + \frac{a_{d-1}}{a_d z} + \frac{a_{d-2}}{a_d z^2} + \cdots + \frac{a_1}{a_d z^{d-1}} + \frac{a_0}{a_d z^d} \right|.$$

Then the sum inside the last factor has limit 1 as $z \to \infty$ (by Exercise 3.12), and so its modulus is between $\tfrac{1}{2}$ and 2 as long as $|z|$ is large enough. □

Theorem 5.11 (Fundamental Theorem of Algebra[2])**.** *Every nonconstant polynomial has a root in $\mathbb{C}$.*

---

[2] The Fundamental Theorem of Algebra was first proved by Gauß (in his doctoral dissertation in 1799, which had a flaw—later, he provided three rigorous proofs), although its statement had been assumed to

Proof. Suppose (by way of contradiction) that p does not have any roots, that is, $p(z) \neq 0$ for all $z \in \mathbb{C}$. Then $\frac{1}{p(z)}$ is entire, and so Cauchy's Integral Formula (Theorem 4.24) gives

$$\frac{1}{p(0)} = \frac{1}{2\pi i} \int_{C[0,R]} \frac{\frac{1}{p(z)}}{z} \, dz,$$

for any $R > 0$. Let d be the degree of $p(z)$ and a_d its leading coefficient. Propositions 4.6((d)) and 5.10 allow us to estimate, for sufficiently large R,

$$\left|\frac{1}{p(0)}\right| = \frac{1}{2\pi}\left|\int_{C[0,R]} \frac{dz}{z\, p(z)}\right| \leq \frac{1}{2\pi} \max_{z \in C[0,R]} \left|\frac{1}{z\, p(z)}\right| 2\pi R \leq \frac{2}{|a_d| R^d}.$$

The left-hand side is independent of R, while the right-hand side can be made arbitrarily small (by choosing R sufficiently large), and so we conclude that $\frac{1}{p(0)} = 0$, which is impossible. □

Theorem 5.11 implies that any polynomial p can be factored into linear terms of the form $z - a$ where a is a root of p, as we can apply the corollary, after getting a root a, to $\frac{p(z)}{z-a}$ (which is again a polynomial by the division algorithm), etc. (see also Exercise 5.11).

A short reformulation of the Fundamental Theorem of Algebra (Theorem 5.11) is to say that $\mathbb{C}$ is *algebraically closed*. In contrast, $\mathbb{R}$ is not algebraically closed.

Example 5.12. The polynomial $p(x) = 2x^4 + 5x^2 + 3$ no roots in $\mathbb{R}$. The Fundamental Theorem of Algebra (Theorem 5.11) states that p must have a root (in fact, four roots) in $\mathbb{C}$:

$$p(x) = (x^2 + 1)(2x^2 + 3) = (x+i)(x-i)(\sqrt{2}x + \sqrt{3}i)(\sqrt{2}x - \sqrt{3}i) \quad □$$

Another powerful consequence of Theorem 5.1 is the following result, which again has no counterpart in real analysis (consider, for example, the real sine function).

---

be correct long before Gauß's time. It is amusing that such an important algebraic result can be proved purely analytically. There are proofs of the Fundamental Theorem of Algebra that do not use complex analysis. On the other hand, all proofs use *some* analysis (such as the Intermediate Value Theorem). The Fundamental Theorem of Algebra refers to *algebra* in the sense that it existed in 1799, not to modern algebra. Thus one might say that the Fundamental Theorem of Algebra is neither fundamental to algebra nor even a theorem of algebra. The proof we give here is due to Anton R. Schep and appeared in the *American Mathematical Monthly* (January 2009).

Corollary 5.13 (Liouville's[3] Theorem). Any bounded entire function is constant.

Proof. Suppose $|f(z)| \leq M$ for all $z \in \mathbb{C}$. Given any $w \in \mathbb{C}$, we apply Theorem 5.1 with the circle $C[w, R]$; note that we can choose any $R > 0$ because f is entire. By Proposition 4.6((d)),

$$|f'(w)| = \left| \frac{1}{2\pi i} \int_{C[w,R]} \frac{f(z)}{(z-w)^2} dz \right| \leq \frac{1}{2\pi} \max_{z \in C[w,R]} \left| \frac{f(z)}{(z-w)^2} \right| 2\pi R$$

$$= \frac{\max_{z \in C[w,R]} |f(z)|}{R} \leq \frac{M}{R}.$$

The right-hand side can be made arbitrarily small, as we are allowed to choose R as large as we want. This implies that $f' = 0$, and hence, by Theorem 2.17, f is constant. □

As an example of the usefulness of Liouville's theorem (Corollary 5.13), we give another proof of the Fundamental Theorem of Algebra, close to Gauß's original proof.

Second proof of the Fundamental Theorem of Algebra (Theorem 5.11). Suppose that p does not have any roots, that is, $p(z) \neq 0$ for all $z \in \mathbb{C}$. Thus the function $f(z) = \frac{1}{p(z)}$ is entire. But $f \to 0$ as $|z| \to \infty$, by Proposition 5.10; consequently, by Exercise 5.10, f is bounded. Now we apply Corollary 5.13 to deduce that f is constant. Hence p is constant, which contradicts our assumptions. □

As one more example of the theme of getting results from Cauchy's Integral Formulas by taking the limit as a path "goes to infinity," we compute an improper integral.

Example 5.14. We will compute the (real) integral

$$\int_{-\infty}^{\infty} \frac{dx}{x^2 + 1} = \pi.$$

Let σ_R be the counterclockwise semicircle formed by the segment $[-R, R]$ of the real axis from $-R$ to R, followed by the circular arc γ_R of radius R in the upper half plane from R to $-R$, where $R > 1$; see Figure 5.2.

---

[3] This theorem is for historical reasons erroneously attributed to Joseph Liouville (1809–1882). It was published earlier by Cauchy; in fact, Gauß may well have known about it before Cauchy.

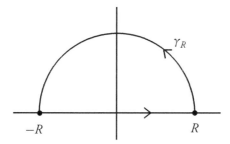

Figure 5.2: The integration paths in Example 5.14.

We computed the integral over σ_R already in Example 4.29:

$$\int_{\sigma_R} \frac{dz}{z^2+1} = \pi.$$

This holds for any $R > 1$, and so we can take the limit as $R \to \infty$. By Proposition 4.6((d)) and the reverse triangle inequality (Corollary 1.7((b))),

$$\left| \int_{\gamma_R} \frac{dz}{z^2+1} \right| \leq \max_{z \in \gamma_R} \left| \frac{1}{z^2+1} \right| \pi R \leq \max_{z \in \gamma_R} \left(\frac{1}{|z|^2 - 1} \right) \pi R = \frac{\pi R}{R^2 - 1}$$

which goes to 0 as $R \to \infty$. Thus

$$\pi = \lim_{R \to \infty} \int_{\sigma_R} \frac{dz}{z^2+1} = \lim_{R \to \infty} \int_{[-R,R]} \frac{dz}{z^2+1} + \lim_{R \to \infty} \int_{\gamma_R} \frac{dz}{z^2+1} = \int_{-\infty}^{\infty} \frac{dx}{x^2+1}.$$

Of course this integral can be evaluated almost as easily using standard formulas from calculus. However, just slight modifications of this example lead to improper integrals that are beyond the scope of basic calculus; see Exercises 5.18 and 5.19. □

Exercises

5.1. Compute the following integrals, where □ is the boundary of the square with vertices at $\pm 4 \pm 4i$, positively oriented:

(a) $\int_\square \frac{\exp(z^2)}{z^3} dz$

(b) $\int_\square \frac{\exp(3z)}{(z-\pi i)^2} dz$

(c) $\int_{\square} \dfrac{\sin(2z)}{(z-\pi)^2}\,dz$ \qquad (d) $\int_{\square} \dfrac{\exp(z)\cos(z)}{(z-\pi)^3}\,dz$

5.2. Prove the formula for f'' in Theorem 5.1.

Hint: Modify the proof of the integral formula for $f'(w)$ as follows:

(a) Write a difference quotient for $f''(w)$, and use the formula for $f'(w)$ in Theorem 5.1 to convert this difference quotient into an integral of $f(z)$ divided by some polynomial.

(b) Subtract the desired integral formula for $f''(w)$ from your integral for the difference quotient, and simplify to get the analogue of (5.1).

(c) Find a bound as in the proof of Theorem 5.1 for the integrand, and conclude that the limit of the difference quotient is the desired integral formula.

5.3. Integrate the following functions over the circle $C[0,3]$:

(a) $\mathrm{Log}(z-4i)$ \qquad (d) $\dfrac{\exp z}{z^3}$ \qquad (g) $\dfrac{\sin z}{(z^2+\tfrac{1}{2})^2}$

(b) $\dfrac{1}{z-\tfrac{1}{2}}$ \qquad (e) $\left(\dfrac{\cos z}{z}\right)^2$ \qquad (h) $\dfrac{1}{(z+4)(z^2+1)}$

(c) $\dfrac{1}{z^2-4}$ \qquad (f) i^{z-3} \qquad (i) $\dfrac{\exp(2z)}{(z-1)^2(z-2)}$

5.4. Compute $\displaystyle\int_{C[0,2]} \dfrac{\exp z}{(z-w)^2}\,dz$ where w is any fixed complex number with $|w|\neq 2$.

5.5. Define $f : D[0,1] \to \mathbb{C}$ through

$$f(z) := \int_{[0,1]} \frac{dw}{1-wz}$$

(the integration path is from 0 to 1 along the real line). Prove that f is holomorphic in the unit disk $D[0,1]$.

5.6. To appreciate Corollary 5.5, show that the function $f : \mathbb{R} \to \mathbb{R}$ given by

$$f(x) := \begin{cases} x^2 \sin(\frac{1}{x}) & \text{if } x \neq 0, \\ 0 & \text{if } x = 0 \end{cases}$$

is differentiable in $\mathbb{R}$, yet f' is not even continuous (much less differentiable) at 0.

5.7. Prove that $f(z) = \bar{z}^2$ does not have an antiderivative in any nonempty region.

5.8. Show that $\exp(\sin z)$ has an antiderivative on $\mathbb{C}$. (What is it?)

5.9. Find a region on which $f(z) = \exp(\frac{1}{z})$ has an antiderivative. (Your region should be as large as you can make it. How does this compare with the real function $f(x) = e^{\frac{1}{x}}$?)

5.10. Suppose f is continuous on $\mathbb{C}$ and $\lim_{z \to \infty} f(z)$ is finite. Show that f is bounded. (*Hint*: If $\lim_{z \to \infty} f(z) = L$, use the definition of the limit at infinity to show that there is $R > 0$ so that $|f(z) - L| < 1$ if $|z| > R$. Now argue that $|f(z)| < |L| + 1$ for $|z| > R$. Use an argument from calculus to show that $|f(z)|$ is bounded for $|z| \leq R$.)

5.11. Let p be a polynomial of degree $n > 0$. Prove that there exist complex numbers $c, z_1, z_2, \ldots, z_k$ and positive integers $j_1, \ldots, j_k$ such that

$$p(z) = c(z - z_1)^{j_1} (z - z_2)^{j_2} \cdots (z - z_k)^{j_k},$$

where $j_1 + \cdots + j_k = n$.

5.12. Show that a polynomial of odd degree with real coefficients must have a real zero. (*Hint*: Use Exercise 1.24.)

5.13. Suppose f is entire and $|f(z)| \leq \sqrt{|z|}$ for all $z \in \mathbb{C}$. Prove that f is constant.

5.14. Suppose f is entire and there exists $M > 0$ such that $|f(z)| \geq M$ for all $z \in \mathbb{C}$. Prove that f is constant.

5.15. Suppose f is entire with bounded real part, i.e., writing $f(z) = u(z) + i\, v(z)$, there exists $M > 0$ such that $|u(z)| \leq M$ for all $z \in \mathbb{C}$. Prove that f is constant. (*Hint*: Consider the function $\exp(f(z))$.)

5.16. Suppose f is entire and there exist constants a and b such that $|f(z)| \leq a|z| + b$ for all $z \in \mathbb{C}$. Prove that f is a polynomial of degree at most 1. (*Hint*: Use Theorem 5.1 and Exercise 2.28.)

5.17. Suppose $f : D[0,1] \to D[0,1]$ is analytic. Prove that for $|z| < 1$,
$$|f'(z)| \leq \frac{1}{1-|z|}.$$

5.18. Compute $\displaystyle\int_{-\infty}^{\infty} \frac{dx}{x^4 + 1}$.

5.19. In this problem $f(z) = \frac{\exp(iz)}{z^2+1}$ and $R > 1$. Modify our computations in Example 5.14 as follows.

(a) Show that $\int_{\sigma_R} f = \frac{\pi}{e}$ where σ_R is again (as in Figure 5.2) the counterclockwise semicircle formed by the segment $[-R, R]$ on the real axis, followed by the circular arc γ_R of radius R in the upper half plane from R to $-R$.

(b) Show that $|\exp(iz)| \leq 1$ for z in the upper half plane, and conclude that $|f(z)| \leq \frac{2}{|z|^2}$ for sufficiently large $|z|$.

(c) Show that $\lim_{R \to \infty} \int_{\gamma_R} f = 0$ and hence $\lim_{R \to \infty} \int_{[-R,R]} f = \frac{\pi}{e}$.

(d) Conclude, by just considering the real part, that
$$\int_{-\infty}^{\infty} \frac{\cos(x)}{x^2 + 1}\, dx = \frac{\pi}{e}.$$

5.20. Compute $\displaystyle\int_{-\infty}^{\infty} \frac{\cos(x)}{x^4 + 1}\, dx$.

5.21. This exercise outlines how to extend some of the results of this chapter to the Riemann sphere as defined in Section 3.2. Suppose $G \subseteq \mathbb{C}$ is a region that contains 0, let f be a continuous function on G, and let $\gamma \subset G \setminus \{0\}$ be a piecewise smooth path in G avoiding the origin, parametrized as $\gamma(t)$, $a \leq t \leq b$.

(a) Show that
$$\int_\gamma f(z)\,dz = \int_\sigma f\left(\frac{1}{z}\right)\frac{1}{z^2}\,dz$$
where $\sigma(t) := \frac{1}{\gamma(t)}$, $a \leq t \leq b$.

Now suppose $\lim_{z \to 0} f\left(\frac{1}{z}\right)\frac{1}{z^2} = L$ is finite. Let $H := \{\frac{1}{z} : z \in G \setminus \{0\}\}$ and define the function $g : H \cup \{0\} \to \mathbb{C}$ by

$$g(z) := \begin{cases} f\left(\frac{1}{z}\right)\frac{1}{z^2} & \text{if } z \in H, \\ L & \text{if } z = 0. \end{cases}$$

Thus g is continuous on $H \cup \{0\}$ and (a) gives the identity

$$\int_\gamma f = \int_\sigma g.$$

In particular, we can transfer certain properties between these two integrals. For example, if $\int_\sigma g$ is path independent, so is $\int_\gamma f$. Here is but one application:

(a) Show that $\int_\gamma z^n\,dz$ is path independent for any integer $n \neq -1$.

(b) Conclude (once more) that $\int_\gamma z^n\,dz = 0$ for any integer $n \neq -1$.

Chapter 6

Harmonic Functions

The shortest route between two truths in the real domain passes through the complex domain.
Jacques Hadamard (1865–1963)

We will now spend a short while on certain functions defined on subsets of the complex plane that are *real* valued, namely those functions that are harmonic in some region. The main motivation for studying harmonic functions is that the partial differential equation they satisfy is very common in the physical sciences. Their definition briefly showed its face in Chapter 2, but we study them only now in more detail, since we have more machinery at our disposal. This machinery comes from *complex*-valued functions, which are, nevertheless, intimately connected to harmonic functions.

6.1 Definition and Basic Properties

Recall from Section 2.3 the definition of a harmonic function:

Definition. Let $G \subseteq \mathbb{C}$ be a region. A function $u : G \to \mathbb{R}$ is *harmonic* in G if it has continuous second partials in G and satisfies the *Laplace*[1] *equation*

$$u_{xx} + u_{yy} = 0.$$

Example 6.1. The function $u(x, y) = xy$ is harmonic in $\mathbb{C}$ since $u_{xx} + u_{yy} = 0 + 0 = 0$. □

Example 6.2. The function $u(x, y) = e^x \cos(y)$ is harmonic in $\mathbb{C}$ because

$$u_{xx} + u_{yy} = e^x \cos(y) - e^x \cos(y) = 0.$$
□

---
[1] Named after Pierre-Simon Laplace (1749–1827).

There are (at least) two reasons why harmonic functions are part of the study of complex analysis, and they can be found in the next two theorems.

Proposition 6.3. Suppose $f = u + iv$ is holomorphic in the region G. Then u and v are harmonic in G.

Proof. First, by Corollary 5.5, u and v have continuous second partials. By Theorem 2.13, u and v satisfy the Cauchy–Riemann equations (2.3)

$$u_x = v_y \quad \text{and} \quad u_y = -v_x$$

in G. Hence we can repeat our argumentation in (2.4),

$$u_{xx} + u_{yy} = (u_x)_x + (u_y)_y = (v_y)_x + (-v_x)_y = v_{yx} - v_{xy} = 0.$$

Note that in the last step we used the fact that v has continuous second partials. The proof that v satisfies the Laplace equation is practically identical. □

Proposition 6.3 gives us an effective way to show that certain functions are harmonic in G by way of constructing an accompanying holomorphic function on G.

Example 6.4. Revisiting Example 6.1, we can see that $u(x, y) = xy$ is harmonic in $\mathbb{C}$ also by noticing that

$$f(z) = \tfrac{1}{2} z^2 = \tfrac{1}{2}(x^2 - y^2) + ixy$$

is entire and $\operatorname{Im}(f) = u$. □

Example 6.5. A second reason that the function $u(x, y) = e^x \cos(y)$ from Example 6.2 is harmonic in $\mathbb{C}$ is that

$$f(z) = \exp(z) = e^x \cos(y) + i e^x \sin(y)$$

is entire and $\operatorname{Re}(f) = u$. □

Proposition 6.3 practically shouts for a converse. There are, however, functions that are harmonic in a region G but not the real part (say) of a holomorphic function in G (Exercise 6.5). We do obtain a converse of Proposition 6.3 if we restrict ourselves to *simply-connected* regions.

DEFINITION AND BASIC PROPERTIES

Theorem 6.6. Suppose u is harmonic on a simply-connected region G. Then there exists a harmonic function v in G such that $f = u + iv$ is holomorphic in G.

The function v is called a *harmonic conjugate* of u.

Proof. We will explicitly construct a holomorphic function f (and thus $v = \operatorname{Im} f$). First, let

$$g := u_x - i u_y.$$

The plan is to prove that g is holomorphic, and then to construct an antiderivative of g, which will be almost the function f that we're after. To prove that g is holomorphic, we use Theorem 2.13: first because u is harmonic, $\operatorname{Re} g = u_x$ and $\operatorname{Im} g = -u_y$ have continuous partials. Moreover, again because u is harmonic, $\operatorname{Re} g$ and $\operatorname{Im} g$ satisfy the Cauchy–Riemann equations (2.3):

$$(\operatorname{Re} g)_x = u_{xx} = -u_{yy} = (\operatorname{Im} g)_y$$

and

$$(\operatorname{Re} g)_y = u_{xy} = u_{yx} = -(\operatorname{Im} g)_x.$$

Theorem 2.13 implies that g is holomorphic in G, and so we can use Corollary 5.8 to obtain an antiderivative h of g on G (here is where we use the fact that G is simply connected). Now we decompose h into its real and imaginary parts as $h = a + ib$. Then, again using Theorem 2.13,

$$g = h' = a_x + i b_x = a_x - i a_y.$$

(The second equation follows from the Cauchy–Riemann equations (2.3).) But the real part of g is u_x, so we obtain $u_x = a_x$ and thus $u(x, y) = a(x, y) + c(y)$ for some function c that depends only on y. On the other hand, comparing the imaginary parts of g and h' yields $-u_y = -a_y$ and so $u(x, y) = a(x, y) + c(x)$ where c depends only on x. Hence c has to be constant, and $u(x, y) = a(x, y) + c$. But then

$$f(z) := h(z) + c$$

is a function holomorphic in G whose real part is u, as promised. □

As a side remark, with hindsight it should not be surprising that the function g that we first constructed in our proof is the derivative of the sought-after function

f. Namely, by Theorem 2.13 such a function $f = u + iv$ must satisfy

$$f' = u_x + i v_x = u_x - i u_y.$$

(The second equation follows from the Cauchy–Riemann equations (2.3).) It is also worth mentioning that our proof of Theorem 6.6 shows that if u is harmonic in G, then u_x is the real part of the function $g = u_x - i u_y$, which is holomorphic in G regardless of whether G is simply connected or not.

As our proof of Theorem 6.6 is constructive, we can use it to produce harmonic conjugates.

Example 6.7. Revisiting Example 6.1 for the second time, we can construct a harmonic conjugate of $u(x, y) = xy$ along the lines of our proof of Theorem 6.6: first let

$$g := u_x - i u_y = y - i x = -i z$$

which has antiderivative

$$h(z) = -\tfrac{i}{2} z^2 = xy - \tfrac{i}{2}(x^2 - y^2)$$

whose real part is u and whose imaginary part

$$v(x, y) := -\tfrac{1}{2}(x^2 - y^2)$$

gives a harmonic conjugate for u. □

We can give a more practical machinery for computing harmonic conjugates, which only depends on computing certain (calculus) integrals; thus this can be easily applied, e.g., to polynomials. We state it for functions that are harmonic in the whole complex plane, but you can easily adjust it to functions that are harmonic on certain subsets of $\mathbb{C}$.[2]

Theorem 6.8. Suppose u is harmonic on $\mathbb{C}$. Then

$$v(x, y) := \int_0^y \frac{\partial u}{\partial x}(x, t)\, dt - \int_0^x \frac{\partial u}{\partial y}(t, 0)\, dt$$

is a harmonic conjugate for u.

---

[2] Theorem 6.8 is due to Sheldon Axler and the basis for his Mathematica package Harmonic Function Theory.

Proof. We will prove that $u + iv$ satisfies the Cauchy–Riemann equations (2.3). The first follows from

$$\frac{\partial v}{\partial y}(x, y) = \frac{\partial u}{\partial x}(x, y),$$

by the Fundamental Theorem of Calculus (Theorem A.3).

Second, by Leibniz's Rule (Theorem A.9), the Fundamental Theorem of Calculus (Theorem A.3), and the fact that u is harmonic,

$$\frac{\partial v}{\partial x}(x, y) = \int_0^y \frac{\partial^2 u}{\partial x^2}(x, t) \, dt - \frac{\partial u}{\partial y}(x, 0) = -\int_0^y \frac{\partial^2 u}{\partial t^2}(x, t) \, dt - \frac{\partial u}{\partial y}(x, 0)$$
$$= -\left(\frac{\partial u}{\partial y}(x, y) - \frac{\partial u}{\partial y}(x, 0) \right) - \frac{\partial u}{\partial y}(x, 0) = -\frac{\partial u}{\partial y}(x, y). \qquad \square$$

As you might imagine, Proposition 6.3 and Theorem 6.6 allow for a powerful interplay between harmonic and holomorphic functions. In that spirit, the following theorem appears not too surprising. You might appreciate its depth better when looking back at the simple definition of a harmonic function.

Corollary 6.9. *A harmonic function is infinitely differentiable.*

Proof. Suppose u is harmonic in G and $z_0 \in G$. We will show that $u^{(n)}(z_0)$ exists for all positive integers n. Let $r > 0$ such that the disk $D[z_0, r]$ is contained in G. Since $D[z_0, r]$ is simply connected, Theorem 6.6 asserts the existence of a holomorphic function f in $D[z_0, r]$ such that $u = \operatorname{Re} f$ on $D[z_0, r]$. By Corollary 5.5, f is infinitely differentiable on $D[z_0, r]$, and hence so is its real part u. $\qquad \square$

This proof is the first in a series of proofs that uses the fact that the property of being harmonic is *local*—it is a property at each point of a certain region. Note that in our proof of Corollary 6.9 we did not construct a function f that is holomorphic in G; we only constructed such a function on the disk $D[z_0, r]$. This f might very well differ from one disk to the next.

6.2 Mean-Value and Maximum/Minimum Principle

We have established an intimate connection between harmonic and holomorphic functions, and so it should come as no surprise that some of the theorems we proved for holomorphic functions have analogues in the world of harmonic functions. Here is such a harmonic analogue of Cauchy's Integral Formula (Theorems 4.24 and 4.27).

Theorem 6.10. Suppose u is harmonic in the region G and $\overline{D}[w, r] \subset G$. Then

$$u(w) = \frac{1}{2\pi} \int_0^{2\pi} u(w + r e^{it}) \, dt \,.$$

Proof. Exercise 6.14 provides R so that $\overline{D}[w, r] \subset D[w, R] \subset G$. The open disk $D[w, R]$ is simply connected, so by Theorem 6.6 there is a function f holomorphic in $D[w, R]$ such that $u = \operatorname{Re} f$ on $D[w, R]$. Now we apply Corollary 4.25 to f:

$$f(w) = \frac{1}{2\pi} \int_0^{2\pi} f(w + r e^{it}) \, dt \,.$$

Theorem 6.10 follows by taking the real part on both sides. □

Corollary 4.25 and Theorem 6.10 say that holomorphic and harmonic functions have the *mean-value property*. Our next result is an important consequence of this property to extreme values of a function.

Definition. Let $G \subset \mathbb{C}$ be a region. The function $u : G \to \mathbb{R}$ has a *strong relative maximum* at $w \in G$ if there exists a disk $D[w, r] \subseteq G$ such that $u(z) \leq u(w)$ for all $z \in D[w, r]$ and $u(z_0) < u(w)$ for some $z_0 \in D[w, r]$. The definition of a *strong relative minimum* is analogous.

Theorem 6.11. If u is harmonic in the region G, then it does not have a strong relative maximum or minimum in G.

Proof. Assume, by way of contradiction, that w is a strong relative maximum. Then there is a disk in G centered at w containing a point z_0 with $u(z_0) < u(w)$. Let $r := |z_0 - w|$ and apply Theorem 6.10:

$$u(w) = \frac{1}{2\pi} \int_0^{2\pi} u(w + r e^{it}) \, dt \,.$$

Intuitively, this cannot hold, because some of the function values we're integrating are smaller than $u(w)$, contradicting the mean-value property. To make this into a thorough argument, suppose that $z_0 = w + r e^{it_0}$ for $0 \leq t_0 < 2\pi$. Because $u(z_0) < u(w)$ and u is continuous, there is a whole interval of parameters $[t_0, t_1] \subseteq [0, 2\pi]$ such that $u(w + r e^{it}) < u(w)$ for $t_0 \leq t \leq t_1$. Now we split up the mean-value integral:

$$u(w) = \frac{1}{2\pi} \int_0^{2\pi} u(w + r e^{it}) dt$$
$$= \frac{1}{2\pi} \left(\int_0^{t_0} u(w + r e^{it}) dt + \int_{t_0}^{t_1} u(w + r e^{it}) dt + \int_{t_1}^{2\pi} u(w + r e^{it}) dt \right)$$

All the integrands can be bounded by $u(w)$; for the middle integral we get a *strict* inequality. Hence

$$u(w) < \frac{1}{2\pi} \left(\int_0^{t_0} u(w) dt + \int_{t_0}^{t_1} u(w) dt + \int_{t_1}^{2\pi} u(w) dt \right) = u(w),$$

a contradiction.

The same argument works if we assume that u has a relative minimum. But in this case there's a shortcut argument: if u has a strong relative minimum then the harmonic function $-u$ has a strong relative maximum, which we just showed cannot exist. □

So far, harmonic functions have benefited from our knowledge of holomorphic functions. Here is a result where the benefit goes in the opposite direction.

Corollary 6.12. *If f is holomorphic and nonzero in the region G, then $|f|$ does not have a strong relative maximum or minimum in G.*

Proof. By Exercise 6.6, the function $\ln|f(z)|$ is harmonic on G and so, by Theorem 6.11, does not have a strong relative maximum or minimum in G. But then neither does $|f(z)|$, because $\ln$ is monotonic. □

We finish our excursion about harmonic functions with a preview and its consequences. We say a real valued function u on a region G has a *weak relative maximum* at w if there exists a disk $D[w, r] \subseteq G$ such that all $z \in D[w, r]$ satisfy $u(z) \le u(w)$. We define *weak relative minimum* similarly. In Chapter 8 we will strengthen Theorem 6.11 and Corollary 6.12 to Theorem 8.17 and Corollary 8.20 by replacing *strong relative extremum* in the hypotheses with *weak relative extremum*.[3] A special but important case of the maximum/minimum principle for harmonic functions,

---

[3] In particular, we will show that one does not have to assume that f is nonzero in a region G to have a strong relative maximum in G.

Corollary 8.20, concerns *bounded* regions. In Chapter 8 we will establish that, if u is harmonic in a bounded region G and continuous on its closure, then

$$\sup_{z \in G} u(z) = \max_{z \in \partial G} u(z) \quad \text{and} \quad \inf_{z \in G} u(z) = \min_{z \in \partial G} u(z) \qquad (6.1)$$

where, as usual, ∂G denotes the boundary of G. We'll exploit this in the next two corollaries.

Corollary 6.13. Suppose u is harmonic in the bounded region G and continuous on its closure. If u is zero on ∂G then u is zero in G.

Proof. By (6.1),
$$u(z) \leq \sup_{z \in G} u(z) = \max_{z \in \partial G} u(z) = 0$$

and
$$u(z) \geq \inf_{z \in G} u(z) = \min_{z \in \partial G} u(z) = 0,$$

so u must be zero in G. □

Corollary 6.14. Suppose u and v are harmonic in the bounded region G and continuous on its closure. If $u(z) = v(z)$ for all $z \in \partial G$ then $u(z) = v(z)$ for all $z \in G$.

Proof. $u - v$ is harmonic in G (Exercise 6.2) and is continuous on the closure $\overline{G}$, and $u - v$ is zero on ∂G. Now apply Corollary 6.13. □

Corollary 6.14 says that if a function u is harmonic in a bounded region G and is continuous on the closure $\overline{G}$ then the values of u at points in G are completely determined by the values of u on the boundary of G. We should remark, however, that this result is of a completely theoretical nature: it says nothing about how to extend a continuous function u given on the boundary of a region to be harmonic in the full region. This problem is called the *Dirichlet*[4] *problem*, and it has a solution for all bounded simply-connected regions. If the region is the unit disk and u is a continuous function on the unit circle, define

$$\hat{u}(e^{i\varphi}) := u(e^{i\varphi}) \quad \text{and} \quad \hat{u}(r e^{i\varphi}) := \frac{1}{2\pi} \int_0^{2\pi} u(e^{it}) P_r(\varphi - t) \, dt \quad \text{for } r < 1,$$

---
[4]Named after Johann Peter Gustav Dirichlet (1805–1859).

where $P_r(\varphi)$ is the Poisson kernel which we introduced in Exercise 4.30. Then $\hat{u}$ is the desired extension: it is continuous on the closed unit disk, harmonic in the open unit disk, and agrees with u on the unit circle. In simple cases this solution can be converted to solutions in other regions, using a conformal map to the unit disk. All of this is beyond the scope of this book, though Exercise 6.13 gives some indication why the above formula does the trick. At any rate, we remark that Corollary 6.14 says that the solution to the Dirichlet problem is unique.

Exercises

6.1. Show that all partial derivatives of a harmonic function are harmonic.

6.2. Suppose $u(x,y)$ and $v(x,y)$ are harmonic in G, and $c \in \mathbb{R}$. Prove that $u(x,y) + c\,v(x,y)$ is also harmonic in G.

6.3. Give an example that shows that the product of two harmonic functions is not necessarily harmonic.

6.4. Let $u(x,y) = e^x \sin y$.

  (a) Show that u is harmonic on $\mathbb{C}$.

  (b) Find an entire function f such that $\text{Re}(f) = u$.

6.5. Consider $u(x,y) = \ln(x^2 + y^2)$.

  (a) Show that u is harmonic on $\mathbb{C} \setminus \{0\}$.

  (b) Prove that u is *not* the real part of a function that is holomorphic in $\mathbb{C} \setminus \{0\}$.

6.6. Show that, if f is holomorphic and nonzero in G, then $\ln|f(x,y)|$ is harmonic in G.

6.7. Suppose $u(x,y)$ is a function $\mathbb{R}^2 \to \mathbb{R}$ that depends only on x. When is u harmonic?

6.8. Is it possible to find a real function $v(x,y)$ so that $x^3 + y^3 + i\,v(x,y)$ is holomorphic?

6.9. Suppose f is holomorphic in the region $G \subseteq \mathbb{C}$ with image $H := \{f(z) : z \in G\}$, and u is harmonic on H. Show that $u(f(z))$ is harmonic on G.

6.10. Suppose $u(r,\varphi)$ is a function $\mathbb{R}^2 \to \mathbb{R}$ given in terms of polar coordinates.

(a) Show that the Laplace equation for $u(r,\varphi)$ is

$$\frac{1}{r}u_r + u_{rr} + \frac{1}{r^2}u_{\varphi\varphi} = 0.$$

(b) Show that $u(r,\varphi) = r^2 \cos(2\varphi)$ is harmonic on $\mathbb{C}$. Generalize.

(c) If $u(r,\varphi)$ depends only on r, when is u harmonic?

(d) If $u(r,\varphi)$ depends only on φ, when is u harmonic?

6.11. Prove that, if u is harmonic and bounded on $\mathbb{C}$, then u is constant. (*Hint*: Use Theorem 6.6 and Liouville's Theorem (Corollary 5.13).)

6.12. Suppose $u(x,y)$ is a harmonic polynomial in x and y. Prove that the harmonic conjugate of u is also a polynomial in x and y.

6.13. Recall from Exercise 4.30 the Poisson kernel

$$P_r(\varphi) = \frac{1-r^2}{1-2r\cos(\varphi)+r^2},$$

where $0 < r < 1$. In this exercise, we will prove the *Poisson Integral Formula*: if u is harmonic on an open set containing the closed unit disk $\overline{D}[0,1]$ then for any $r < 1$

$$u(r\,e^{i\varphi}) = \frac{1}{2\pi}\int_0^{2\pi} u(e^{it})\,P_r(\varphi - t)\,dt. \tag{6.2}$$

Suppose u is harmonic on an open set containing $\overline{D}[0,1]$. By Exercise 6.14 we can find $R_0 > 1$ so that u is harmonic in $D[0,R_0]$.

(a) Recall the Möbius function

$$f_a(z) = \frac{z-a}{1-\overline{a}z},$$

for some fixed $a \in \mathbb{C}$ with $|a| < 1$, from Exercise 3.9. Show that $u(f_{-a}(z))$ is harmonic on an open disk $D[0, R_1]$ containing $\overline{D}[0, 1]$.

(b) Apply Theorem 6.10 to the function $u(f_{-a}(z))$ with $w = 0$ to deduce

$$u(a) = \frac{1}{2\pi i} \int_{C[0,1]} \frac{u(f_{-a}(z))}{z} \, dz. \tag{6.3}$$

(c) Recalling, again from Exercise 3.9, that $f_a(z)$ maps the unit circle to itself, apply a change of variables to (6.3) to prove

$$u(a) = \frac{1}{2\pi} \int_0^{2\pi} u(e^{it}) \frac{1 - |a|^2}{|e^{it} - a|^2} \, dt.$$

(d) Deduce (6.2) by setting $a = r e^{i\varphi}$.

6.14. Suppose G is open and $\overline{D}[a, r] \subset G$. Show that there is $R > r$ so that $\overline{D}[a, r] \subset D[a, R] \subset G$. (*Hint:* If $G = \mathbb{C}$ just take $R = r + 1$. Otherwise choose some $w \in \mathbb{C} \setminus G$, let $M = |w - a|$, and let $K = \overline{D}[a, M] \setminus G$. Show that K is nonempty, closed and bounded, and apply Theorem A.1 to find a point $z_0 \in K$ that minimizes $f(z) = |z - a|$ on K. Show that $R = |z_0 - a|$ works.)

Chapter 7

Power Series

It is a pain to think about convergence but sometimes you really have to.
Sinai Robins

Looking back to what machinery we have established so far for integrating complex functions, there are several useful theorems we developed in Chapters 4 and 5. But there are some simple-looking integrals, such as

$$\int_{C[2,3]} \frac{\exp(z)}{\sin(z)} \, dz, \qquad (7.1)$$

that we cannot compute with this machinery. The problems, naturally, comes from

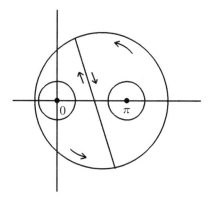

Figure 7.1: Modifying the integration path for (7.1).

the singularities at 0 and π inside the integration path, which in turn stem from the roots of the sine function. We might try to simplify this problem a bit by writing the integral as the sum of integrals over the two "D" shaped paths shown in Figure 5.1

(the integrals along the common straight line segments cancel). Furthermore, by Cauchy's Theorem 4.18, we may replace these integrals with integrals over small circles around 0 and π. This transforms (7.1) into a sum of two integrals, which we are no closer to being able to compute; however, we have localized the problem, in the sense that we now "only" have to compute integrals around *one* of the singularities of our integrand.

This motivates developing techniques to approximate complex functions locally, in analogy with the development of Taylor series in calculus. It is clear that we need to go further here, as we'd like to have such approximations near a singularity of a function. At any rate, to get any of this started, we need to talk about sequences and series of complex numbers and functions, and this chapter develops them.

7.1 Sequences and Completeness

As in the real case,[1] a *(complex) sequence* is a function from the positive (sometimes the nonnegative) integers to the complex numbers. Its values are usually written as a_n (as opposed to $a(n)$) and we commonly denote the sequence by $(a_n)_{n=1}^{\infty}$, $(a_n)_{n \geq 1}$, or simply (a_n). Considering such a sequence as a function of n, the notion of convergence is merely a repeat of the definition we gave in Section 3.2, adjusted to the fact that n is an integer.

Definition. Suppose (a_n) is a sequence and $L \in \mathbb{C}$ such that for all $\varepsilon > 0$ there is an integer N such that for all $n \geq N$, we have $|a_n - L| < \varepsilon$. Then the sequence (a_n) is *convergent* and L is its *limit*; in symbols we write

$$\lim_{n \to \infty} a_n = L.$$

If no such L exists then the sequence (a_n) is *divergent*.

As in our previous definitions of limit, the limit of a sequence is unique if it exists. See Exercise 7.7.

Example 7.1. We claim that $\lim_{n \to \infty} \frac{i^n}{n} = 0$: Given $\varepsilon > 0$, choose $N > \frac{1}{\varepsilon}$. Then for any $n \geq N$,

$$\left| \frac{i^n}{n} - 0 \right| = \left| \frac{i^n}{n} \right| = \frac{|i|^n}{n} = \frac{1}{n} \leq \frac{1}{N} < \varepsilon. \qquad \square$$

---
[1] There will be no surprises in this chapter of the nature *real versus complex*.

To prove that a sequence (a_n) is divergent, we have to show the negation of the statement that defines convergence, that is: given any $L \in \mathbb{C}$, there exists $\varepsilon > 0$ such that, given any integer N, there exists an integer n such that $|a_n - L| \geq \varepsilon$. (If you have not negated many mathematical statements, this is worth meditating about.)

Example 7.2. The sequence $(a_n = i^n)$ diverges: Given $L \in \mathbb{C}$, choose $\varepsilon = \frac{1}{2}$. We consider two cases: If $\operatorname{Re}(L) \geq 0$, then for any N, choose $n \geq N$ such that $a_n = -1$. (This is always possible since $a_{4k+2} = i^{4k+2} = -1$ for any $k \geq 0$.) Then

$$|a_n - L| = |1 + L| \geq 1 > \frac{1}{2}.$$

If $\operatorname{Re}(L) < 0$, then for any N, choose $n \geq N$ such that $a_n = 1$. (This is always possible since $a_{4k} = i^{4k} = 1$ for any $k > 0$.) Then

$$|a_n - L| = |1 - L| > 1 > \frac{1}{2}.$$

This proves that $(a_n = i^n)$ diverges. □

The following limit laws are the cousins of the identities in Propositions 2.4 and 2.6, with one little twist.

Proposition 7.3. Let (a_n) and (b_n) be convergent sequences and $c \in \mathbb{C}$. Then

(a) $\lim_{n\to\infty} a_n + c \lim_{n\to\infty} b_n = \lim_{n\to\infty} (a_n + c\, b_n)$

(b) $\lim_{n\to\infty} a_n \cdot \lim_{n\to\infty} b_n = \lim_{n\to\infty} (a_n \cdot b_n)$

(c) $\dfrac{\lim_{n\to\infty} a_n}{\lim_{n\to\infty} b_n} = \lim_{n\to\infty} \left(\dfrac{a_n}{b_n}\right)$

(d) $\lim_{n\to\infty} a_n = \lim_{n\to\infty} a_{n+1}$

where in (c) we also require that $\lim_{n\to\infty} b_n \neq 0$. Furthermore, if $f : G \to \mathbb{C}$ is continuous at $L := \lim_{n\to\infty} a_n$ and all $a_n \in G$, then

$$\lim_{n\to\infty} f(a_n) = f(L).$$

Again, the proof of this proposition is essentially a repeat from arguments we have given in Chapters 2 and 3, as you should convince yourself in Exercise 7.4.

SEQUENCES AND COMPLETENESS

We will assume, as an axiom, that $\mathbb{R}$ is *complete*. To phrase this precisely, we need the following.

Definition. The sequence (a_n) is *monotone* if it is either nondecreasing ($a_{n+1} \geq a_n$ for all n) or nonincreasing ($a_{n+1} \leq a_n$ for all n).

There are many equivalent ways of formulating the completeness property for the reals. Here is what we'll go by:

Axiom (Monotone Sequence Property)**.** Any bounded monotone sequence converges.

This axiom (or one of its many equivalent statements) gives arguably the most important property of the real number system; namely, that we can, in many cases, determine that a given sequence converges *without knowing the value of the limit*. In this sense we can use the sequence to define a real number.

Example 7.4. Consider the sequence (a_n) defined by

$$a_n := 1 + \frac{1}{2} + \frac{1}{6} + \cdots + \frac{1}{n!}.$$

This sequence is increasing (by definition) and each $a_n \leq 3$ by Exercise 7.9. By the Monotone Sequence Property, (a_n) converges, which allows us to define one of the most famous numbers in all of mathematics,

$$e := 1 + \lim_{n \to \infty} a_n. \qquad \square$$

Example 7.5. Fix $0 \leq r < 1$. We claim that $\lim_{n \to \infty} r^n = 0$: First, the sequence $(a_n = r^n)$ converges because it is decreasing and bounded below by 0. Let $L := \lim_{n \to \infty} r^n$. By Proposition 7.3,

$$L = \lim_{n \to \infty} r^n = \lim_{n \to \infty} r^{n+1} = r \lim_{n \to \infty} r^n = r L.$$

Thus $(1 - r)L = 0$, and so (since $1 - r \neq 0$) we conclude that $L = 0$. $\qquad \square$

We remark that the Monotone Sequence Property implies the *Least Upper Bound Property*: every nonempty set of real numbers with an upper bound has a *least* upper bound. The Least Upper Bound Property, in turn, implies the following theorem, which is often listed as a separate axiom.[2]

Theorem 7.6. [Archimedean[3] Property] *If x is any real number then there is an integer N that is greater than x.*

For a proof see Exercise 7.10. Theorem 7.6 essentially says that infinity is not part of the real numbers. Note that we already used Theorem 7.6 in Example 7.1. The Archimedean Property underlies the construction of an infinite decimal expansion for any real number, while the Monotone Sequence Property shows that any such infinite decimal expansion actually converges to a real number.

We close this discussion of limits with a pair of standard limits. The first of these can be established by calculus methods (such as L'Hôpital's rule (Theorem A.11), by treating n as the variable); both of them can be proved by more elementary considerations. Either way, we leave the proof of the following to Exercise 7.11.

Proposition 7.7. (a) *Exponentials beat polynomials: for any polynomial $p(n)$ (with complex coefficients) and any $c \in \mathbb{C}$ with $|c| > 1$,*

$$\lim_{n \to \infty} \frac{p(n)}{c^n} = 0.$$

(b) *Factorials beat exponentials: for any $c \in \mathbb{C}$,*

$$\lim_{n \to \infty} \frac{c^n}{n!} = 0.$$

7.2 Series

Definition. A *series* is a sequence (a_n) whose members are of the form $a_n = \sum_{k=1}^{n} b_k$ (or $a_n = \sum_{k=0}^{n} b_k$); we call (b_k) the *sequence of terms* of the series. The $a_n = \sum_{k=1}^{n} b_k$ (or $a_n = \sum_{k=0}^{n} b_k$) are the *partial sums* of the series.

If we wanted to be lazy we would define convergence of a series simply by referring to convergence of the partial sums of the series—after all, we just defined

---

[2] Both the Archimedean Property and the Least Upper Bound Property can be used in (different) axiomatic developments of $\mathbb{R}$.

[3] Archimedes of Syracuse (287–212 BCE) attributes this property to Eudoxus of Cnidus (408–355 BCE).

series through sequences. However, there are some convergence features that take on special appearances for series, so we mention them here explicitly. For starters, a series *converges* to the *limit* (or *sum*) L by definition if

$$\lim_{n \to \infty} a_n = \lim_{n \to \infty} \sum_{k=1}^{n} b_k = L.$$

To prove that a series converges we use the definition of limit of a sequence: for any $\varepsilon > 0$ we have to find an N such that for all $n \geq N$,

$$\left| \sum_{k=1}^{n} b_k - L \right| < \varepsilon.$$

In the case of a convergent series, we usually write its limit as $L = \sum_{k=1}^{\infty} b_k$ or $L = \sum_{k \geq 1} b_k$.

Example 7.8. Fix $z \in \mathbb{C}$ with $|z| < 1$. We claim that the *geometric series* $\sum_{k \geq 1} z^k$ converges with limit

$$\sum_{k \geq 1} z^k = \frac{z}{1-z}.$$

In this case, we can compute the partial sums explicitly:

$$\sum_{k=1}^{n} z^k = z + z^2 + \cdots + z^n = \frac{z - z^{n+1}}{1 - z},$$

whose limit as $n \to \infty$ exists by Example 7.5, because $|z| < 1$. □

Example 7.9. Another series whose limit we can compute by manipulating the partial sums is

$$\sum_{k \geq 1} \frac{1}{k^2 + k} = \lim_{n \to \infty} \sum_{k=1}^{n} \left(\frac{1}{k} - \frac{1}{k+1} \right)$$
$$= \lim_{n \to \infty} \left(1 - \frac{1}{2} + \frac{1}{2} - \frac{1}{3} + \frac{1}{3} - \frac{1}{4} + \cdots + \frac{1}{n} - \frac{1}{n+1} \right)$$
$$= \lim_{n \to \infty} \left(1 - \frac{1}{n+1} \right) = 1.$$

A series where most of the terms cancel like this is called *telescoping*. □

Most of the time we can use the completeness property to check convergence of a series, and it is fortunate that the Monotone Sequence Property has a convenient translation into the language of series of real numbers. The partial sums of a series form a nondecreasing sequence if the terms of the series are nonnegative, and this observation immediately yields the following:

Corollary 7.10. If $b_k \in \mathbb{R}_{\geq 0}$ then $\sum_{k\geq 1} b_k$ converges if and only if the partial sums are bounded.

Example 7.11. With this new terminology, we can revisit Example 7.4: Let $b_k = \frac{1}{k!}$. In Example 7.4 we showed that the partial sums

$$\sum_{k=1}^{n} b_k = \sum_{k=1}^{n} \frac{1}{k!}$$

are bounded, and $\sum_{k\geq 1} \frac{1}{k!} = e - 1$. □

Although Corollary 7.10 is a mere direct consequence of the completeness property of $\mathbb{R}$, it is surprisingly useful. Here is one application, sometimes called the *Comparison Test*:

Corollary 7.12. If $b_k \geq c_k \geq 0$ for all $k \geq 1$ and $\sum_{k\geq 1} b_k$ converges then so does $\sum_{k\geq 1} c_k$.

Proof. By Corollary 7.10, the partial sums $\sum_{k=1}^{n} b_k$ are bounded, and thus so are

$$\sum_{k=1}^{n} c_k \leq \sum_{k=1}^{n} b_k.$$

But this means, again by Corollary 7.10, that $\sum_{k\geq 1} c_k$ converges. □

Proposition 7.13. If $\sum_{k\geq 1} b_k$ converges then $\lim_{n\to\infty} b_n = 0$.

The contrapositive of this proposition is often used, sometimes called the *Test for Divergence*:

Corollary 7.14. If $\lim_{n\to\infty} b_n \neq 0$ or $\lim_{n\to\infty} b_n$ does not exist, then $\sum_{k\geq 1} b_k$ diverges.

Example 7.15. Continuing Example 7.8, for $|z| \geq 1$ the geometric series $\sum_{k\geq 1} z^k$ diverges since in this case $\lim_{n\to\infty} z^n$ either does not exist or is not 0. □

Proof of Proposition 7.13. Suppose $\sum_{k\geq 1} b_k$ converges. Then, using Proposition 7.3(a) & (d),

$$0 = \lim_{n\to\infty} \sum_{k=1}^{n} b_k - \lim_{n\to\infty} \sum_{k=1}^{n-1} b_k = \lim_{n\to\infty}\left(\sum_{k=1}^{n} b_k - \sum_{k=1}^{n-1} b_k\right) = \lim_{n\to\infty} b_n. \quad \square$$

A common mistake is to try to use the converse of Proposition 7.13, but the converse is false:

Example 7.16. The *harmonic series* $\sum_{k\geq 1} \frac{1}{k}$ diverges (even though the terms go to 0): If we assume the series converges to L, then

$$\begin{aligned}
L &= 1 + \frac{1}{2} + \frac{1}{3} + \frac{1}{4} + \frac{1}{5} + \frac{1}{6} + \cdots \\
&> \frac{1}{2} + \frac{1}{2} + \frac{1}{4} + \frac{1}{4} + \frac{1}{6} + \frac{1}{6} + \cdots \\
&= 1 + \frac{1}{2} + \frac{1}{3} + \cdots \\
&= L,
\end{aligned}$$

a contradiction. $\qquad\square$

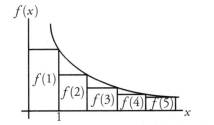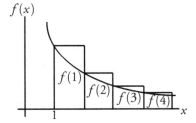

Figure 7.2: The integral test.

Proposition 7.17 (Integral Test). If $f : [1, \infty) \to \mathbb{R}_{\geq 0}$ is continuous and nonincreasing, then

$$\int_1^\infty f(t)\,dt \leq \sum_{k \geq 1} f(k) \leq f(1) + \int_1^\infty f(t)\,dt.$$

The Integral Test literally comes with a proof by picture—see Figure 7.2: the integral of f on the interval $[k, k+1]$ is bounded between $f(k)$ and $f(k+1)$. Adding the pieces gives the inequalities above for the nth partial sum versus the integrals from 1 to n and from 1 to $n+1$, and the inequality persists in the limit.

Corollary 7.18. If $f : [1, \infty) \to \mathbb{R}_{\geq 0}$ is continuous and nonincreasing, then $\sum_{k \geq 1} f(k)$ converges if and only if $\int_1^\infty f(t)\,dt$ is finite.

Proof. Suppose $\int_1^\infty f(t)\,dt = \infty$. Then the first inequality in Proposition 7.17 implies that the partial sums $\sum_{k=1}^n f(k)$ are unbounded, and so Corollary 7.10 says that $\sum_{k \geq 1} f(k)$ cannot converge.

Conversely, if $\int_1^\infty f(t)\,dt$ is finite then the second inequality in Proposition 7.17 says that the partial sums $\sum_{k=1}^n f(k)$ are bounded; thus, again with Corollary 7.10, we conclude that $\sum_{k \geq 1} f(k)$ converges. □

Example 7.19. The series $\sum_{k \geq 1} \frac{1}{k^p}$ converges for $p > 1$ and diverges for $p < 1$ (and the case $p = 1$ was the subject of Example 7.16) because

$$\int_1^\infty \frac{dx}{x^p} = \lim_{a \to \infty} \frac{a^{-p+1}}{-p+1} + \frac{1}{p-1}$$

is finite if and only if $p > 1$. □

By now you might be amused that we have collected several results on series whose terms are nonnegative real numbers. One reason is that such series are a bit easier to handle, another one is that there is a notion of convergence special to series that relates any series to one with only nonnegative terms:

Definition. The series $\sum_{k \geq 1} b_k$ *converges absolutely* if $\sum_{k \geq 1} |b_k|$ converges.

Theorem 7.20. *If a series converges absolutely then it converges.*

This seems like an obvious statement, but its proof is, nevertheless, nontrivial.

Proof. Suppose $\sum_{k \geq 1} |b_k|$ converges. We first consider the case that each b_k is real. Let

$$b_k^+ := \begin{cases} b_k & \text{if } b_k \geq 0, \\ 0 & \text{otherwise} \end{cases} \quad \text{and} \quad b_k^- := \begin{cases} b_k & \text{if } b_k < 0, \\ 0 & \text{otherwise.} \end{cases}$$

Then $0 \leq b_k^+ \leq |b_k|$ and $0 \leq -b_k^- \leq |b_k|$ for all $k \geq 1$, and so by Corollary 7.12, both

$$\sum_{k \geq 1} b_k^+ \quad \text{and} \quad -\sum_{k \geq 1} b_k^-$$

converge. But then so does

$$\sum_{k \geq 1} b_k = \sum_{k \geq 1} b_k^+ + \sum_{k \geq 1} b_k^-.$$

For the general case $b_k \in \mathbb{C}$, we write each term as $b_k = c_k + i\, d_k$. Since $0 \leq |c_k| \leq |b_k|$ for all $k \geq 1$, Corollary 7.12 implies that $\sum_{k \geq 1} c_k$ converges absolutely, and by an analogous argument, so does $\sum_{k \geq 1} d_k$. But now we can use the first case to deduce that both $\sum_{k \geq 1} c_k$ and $\sum_{k \geq 1} d_k$ converge, and thus so does

$$\sum_{k \geq 1} b_k = \sum_{k \geq 1} c_k + i \sum_{k \geq 1} d_k. \qquad \square$$

Example 7.21. Continuing Example 7.19,

$$\zeta(z) := \sum_{k \geq 1} \frac{1}{k^z}$$

converges for $\mathrm{Re}(z) > 1$, because then (using Exercise 3.49)

$$\sum_{k \geq 1} |k^{-z}| = \sum_{k \geq 1} k^{-\mathrm{Re}(z)}$$

converges. Viewed as a function in z, the series $\zeta(z)$ is the *Riemann zeta function*, an indispensable tool in number theory and many other areas in mathematics and physics.[4] $\qquad \square$

---

[4] The Riemann zeta function is the subject of the arguably most famous open problem in mathematics, the *Riemann hypothesis*. It turns out that $\zeta(z)$ can be extended to a function that is holomorphic on $\mathbb{C} \setminus \{1\}$,

Another common mistake is to try to use the converse of Theorem 7.20, which is also false:

Example 7.22. The *alternating harmonic series* $\sum_{k \geq 1} \frac{(-1)^{k+1}}{k}$ converges:

$$\sum_{k \geq 1} \frac{(-1)^{k+1}}{k} = 1 - \frac{1}{2} + \frac{1}{3} - \frac{1}{4} + \frac{1}{5} - \frac{1}{6} + \cdots$$

$$= \left(1 - \frac{1}{2}\right) + \left(\frac{1}{3} - \frac{1}{4}\right) + \left(\frac{1}{5} - \frac{1}{6}\right) + \cdots$$

(There is a small technical detail to be checked here, since we are effectively ignoring half the partial sums of the original series; see Exercise 7.16.) Since

$$\frac{1}{2k-1} - \frac{1}{2k} = \frac{1}{2k(2k-1)} \leq \frac{1}{(2k-1)^2} \leq \frac{1}{k^2},$$

$\sum_{k \geq 1} \frac{(-1)^{k+1}}{k}$ converges by Corollary 7.12 and Example 7.19.

However, according to Example 7.16, $\sum_{k \geq 1} \frac{(-1)^{k+1}}{k}$ does *not* converge absolutely. □

7.3 Sequences and Series of Functions

The fun starts when we study sequences of functions.

Definition. Let $G \subseteq \mathbb{C}$ and $f_n : G \to \mathbb{C}$ for $n \geq 1$. We say that (f_n) *converges pointwise* to $f : G \to \mathbb{C}$ if for each $z \in G$,

$$\lim_{n \to \infty} f_n(z) = f(z).$$

We say that (f_n) *converges uniformly* to $f : G \to \mathbb{C}$ if for all $\varepsilon > 0$ there is an N such that for all $z \in G$ and for all $n \geq N$

$$|f_n(z) - f(z)| < \varepsilon.$$

Sometimes we want to express that either notion of convergence holds only on a subset H of G, in which case we say that (f_n) converges pointwise/uniformly on H.

---

and the Riemann hypothesis asserts that the roots of this extended function in the strip $0 < \text{Re}(z) < 1$ are all on the *critical line* $\text{Re}(z) = \frac{1}{2}$.

It should be clear that uniform convergence on a set implies pointwise convergence on that set; but the converse is not generally true.

Let's digest these two notions of convergence of a function sequence by describing them using quantifiers; as usual, $\forall$ denotes *for all* and $\exists$ means *there exists*. Pointwise convergence on G says

$$\forall \varepsilon > 0 \ \forall z \in G \ \exists N \ \forall n \geq N \ |f_n(z) - f(z)| < \varepsilon,$$

whereas uniform convergence on G translates into

$$\forall \varepsilon > 0 \ \exists N \ \forall z \in G \ \forall n \geq N \ |f_n(z) - f(z)| < \varepsilon.$$

No big deal — we only exchanged two of the quantifiers. In the first case, N may well depend on z, in the second case we need to find an N that works for all $z \in G$. And this can make all the difference ...

Example 7.23. Let $f_n : D[0, 1] \to \mathbb{C}$ be defined by $f_n(z) = z^n$. We claim that this sequence of functions converges pointwise to $f : D[0, 1] \to \mathbb{C}$ given by $f(z) = 0$. This is immediate for the point $z = 0$. Now given any $\varepsilon > 0$ and $0 < |z| < 1$, choose $N > \frac{\ln(\varepsilon)}{\ln|z|}$. Then for all $n \geq N$,

$$|f_n(z) - f(z)| = |z^n - 0| = |z|^n \leq |z|^N < \varepsilon.$$

(You ought to check carefully that all of our inequalities work the way we claim they do.) □

Example 7.24. Let $f_n : D[0, \frac{1}{2}] \to \mathbb{C}$ be defined by $f_n(z) = z^n$. We claim that this sequence of functions converges uniformly to $f : D[0, \frac{1}{2}] \to \mathbb{C}$ given by $f(z) = 0$. Given any $\varepsilon > 0$ and $|z| < \frac{1}{2}$, choose $N > -\ln(2)\ln(\varepsilon)$. Then for all $n \geq N$,

$$|f_n(z) - f(z)| = |z|^n \leq |z|^N < \left(\tfrac{1}{2}\right)^N < \varepsilon.$$

(Again, you should carefully check our inequalities.) □

The differences between Example 7.23 and Example 7.24 are subtle, and we suggest you meditate over them for a while with a good cup of coffee. You might already suspect that the function sequence in Example 7.23 does *not* converge uniformly, as we will see in a moment.

The first application illustrating the difference between pointwise and uniform convergence says, in essence, that if we have a sequence of functions (f_n) that converges uniformly on G then for all $z_0 \in G$

$$\lim_{n \to \infty} \lim_{z \to z_0} f_n(z) = \lim_{z \to z_0} \lim_{n \to \infty} f_n(z).$$

We will need similar interchanges of limits frequently.

Proposition 7.25. Suppose $G \subset \mathbb{C}$ and $f_n : G \to \mathbb{C}$ is continuous, for each $n \geq 1$. If (f_n) converges uniformly to $f : G \to \mathbb{C}$ then f is continuous.

Proof. Let $z_0 \in G$; we will prove that f is continuous at z_0. By uniform convergence, given $\varepsilon > 0$, there is an N such that for all $z \in G$ and all $n \geq N$

$$|f_n(z) - f(z)| < \frac{\varepsilon}{3}.$$

Now we make use of the continuity of the f_n's. This means that given (the same) $\varepsilon > 0$, there is a $\delta > 0$ such that whenever $|z - z_0| < \delta$,

$$|f_n(z) - f_n(z_0)| < \frac{\varepsilon}{3}.$$

All that's left is putting those two inequalities together: by the triangle inequality (Corollary 1.7(c)),

$$\begin{aligned}|f(z) - f(z_0)| &= |f(z) - f_n(z) + f_n(z) - f_n(z_0) + f_n(z_0) - f(z_0)| \\ &\leq |f(z) - f_n(z)| + |f_n(z) - f_n(z_0)| + |f_n(z_0) - f(z_0)| \\ &< \varepsilon.\end{aligned}$$

This proves that f is continuous at z_0. □

Proposition 7.25 can sometimes give a hint that a function sequence does not converge uniformly.

Example 7.26. We modify Example 7.23 and consider the real function sequence $f_n : [0,1] \to \mathbb{R}$ given by $f_n(x) = x^n$. It converges pointwise to $f : [0,1] \to \mathbb{R}$ given by

$$f(x) = \begin{cases} 0 & \text{if } 0 \leq x < 1, \\ 1 & \text{if } x = 1. \end{cases}$$

As this limiting function is not continuous, the above convergence cannot be uniform. This gives a strong indication that the convergence in Example 7.23 is not uniform either, though this needs a separate proof, as the domain of the functions in Example 7.23 is the unit disk (Exercise 7.20(b)). □

Now that we have established Proposition 7.25 about continuity, we can ask about integration of sequences or series of functions. The next theorem should come as no surprise; however, its consequences (which we will see shortly) are wide ranging.

Proposition 7.27. Suppose $f_n : G \to \mathbb{C}$ is continuous, for $n \geq 1$, (f_n) converges uniformly to $f : G \to \mathbb{C}$, and $\gamma \subseteq G$ is a piecewise smooth path. Then

$$\lim_{n \to \infty} \int_\gamma f_n = \int_\gamma f.$$

Proof. We may assume that γ is not just a point, in which case the proposition holds trivially. Given $\varepsilon > 0$, there exists N such that for all $z \in G$ and all $n \geq N$,

$$|f_n(z) - f(z)| < \frac{\varepsilon}{\text{length}(\gamma)}.$$

With Proposition 4.6((d)) we can thus estimate

$$\left| \int_\gamma f_n - \int_\gamma f \right| = \left| \int_\gamma f_n - f \right| \leq \max_{z \in \gamma} |f_n(z) - f(z)| \cdot \text{length}(\gamma) < \varepsilon. \quad □$$

All of these notions for sequences of functions hold verbatim for series of functions. For example, if $\sum_{k \geq 1} f_k(z)$ converges uniformly on G and $\gamma \subseteq G$ is a piecewise smooth path, then

$$\int_\gamma \sum_{k \geq 1} f_k(z) \, dz = \sum_{k \geq 1} \int_\gamma f_k(z) \, dz.$$

In some sense, the above identity is *the* reason we care about uniform convergence.

There are several criteria for uniform convergence; see, e.g., Exercises 7.19 and 7.20, and the following result, sometimes called the *Weierstraß M-test*.

Proposition 7.28. Suppose $f_k : G \to \mathbb{C}$ for $k \geq 1$, and $|f_k(z)| \leq M_k$ for all $z \in G$, where $\sum_{k \geq 1} M_k$ converges. Then $\sum_{k \geq 1} |f_k|$ and $\sum_{k \geq 1} f_k$ converge uniformly in G. (We say the series $\sum_{k \geq 1} f_k$ *converges absolutely and uniformly.*)

Proof. For each fixed z, the series $\sum_{k \geq 1} f_k(z)$ converges absolutely by Corollary 7.12. To show that the convergence is uniform, let $\varepsilon > 0$. Then there exists N such that for all $n \geq N$,

$$\sum_{k \geq 1} M_k - \sum_{k=1}^{n} M_k = \sum_{k > n} M_k < \varepsilon.$$

Thus for all $z \in G$ and $n \geq N$,

$$\left| \sum_{k \geq 1} f_k(z) - \sum_{k=1}^{n} f_k(z) \right| = \left| \sum_{k > n} f_k(z) \right| \leq \sum_{k > n} |f_k(z)| \leq \sum_{k > n} M_k < \varepsilon,$$

which proves uniform convergence. Replace f_k with $|f_k|$ in this argument to see that $\sum_{k \geq 1} |f_k|$ also converges uniformly. □

Example 7.29. We revisit Example 7.8 and consider the geometric series $\sum_{k \geq 1} z^k$ as a series of functions in z. We know from Example 7.8 that this function series converges pointwise for $|z| < 1$:

$$\sum_{k \geq 1} z^k = \frac{z}{1-z}.$$

To study uniform convergence, we apply Proposition 7.28 with $f_k(z) = z^k$. We need a series of upper bounds that converges, so fix a real number $0 < r < 1$ and let $M_k = r^k$. Then

$$|f_k(z)| = |z|^k \leq r^k \qquad \text{for } |z| \leq r,$$

and $\sum_{k \geq 1} r^k$ converges by Example 7.8. Thus, Proposition 7.28 says that $\sum_{k \geq 1} z^k$ converges uniformly for $|z| \leq r$.

We note the subtle distinction of domains for pointwise/uniform convergence: $\sum_{k \geq 1} z^k$ converges (absolutely) for $|z| < 1$, but to force *uniform* convergence, we need to shrink the domain to $|z| \leq r$ for some (arbitrary but fixed) $r < 1$. □

7.4 Regions of Convergence

For the remainder of this chapter (indeed, this book) we concentrate on some very special series of functions.

Definition. A *power series centered at* z_0 is a series of the form

$$\sum_{k \geq 0} c_k (z - z_0)^k$$

where $c_0, c_1, c_2, \ldots \in \mathbb{C}$.

Example 7.30. A slight modification of Example 7.29 gives a fundamental power series, namely the *geometric series*

$$\sum_{k \geq 0} z^k = \frac{1}{1-z}.$$

So here $z_0 = 0$ and $c_k = 1$ for all $k \geq 0$. We note that, as in Example 7.29, this power series converges absolutely for $|z| < 1$ and uniformly for $|z| \leq r$, for any fixed $r < 1$. Finally, as in Example 7.15, the geometric series $\sum_{k \geq 0} z^k$ diverges for $|z| \geq 1$. □

A general power series has a very similar convergence behavior which, in fact, comes from comparing it to a geometric series.

Theorem 7.31. Given a power series $\sum_{k \geq 0} c_k (z - z_0)^k$, there exists a real number $R \geq 0$ or $R = \infty$, such that

(a) $\sum_{k \geq 0} c_k (z - z_0)^k$ converges absolutely for $|z - z_0| < R$;

(b) $\sum_{k \geq 0} c_k (z - z_0)^k$ converges absolutely and uniformly for $|z - z_0| \leq r$, for any $r < R$;

(c) $\sum_{k \geq 0} c_k (z - z_0)^k$ diverges for $|z - z_0| > R$.

We remark that this theorem says nothing about the convergence/divergence of $\sum_{k \geq 0} c_k (z - z_0)^k$ for $|z - z_0| = R$.

Definition. The number R in Theorem 7.31 is called the *radius of convergence* of $\sum_{k \geq 0} c_k (z - z_0)^k$. The open disk $D[z_0, R]$ in which the power series converges absolutely is the *region of convergence*. (If $R = \infty$ then this is $\mathbb{C}$.)

In preparation for the proof of Theorem 7.31, we start with the following observation.

Proposition 7.32. If $\sum_{k \geq 0} c_k (w - z_0)^k$ converges then $\sum_{k \geq 0} c_k (z - z_0)^k$ converges absolutely whenever $|z - z_0| < |w - z_0|$.

Proof. Let $r := |w - z_0|$. If $\sum_{k \geq 0} c_k (w - z_0)^k$ converges then $\lim_{k \to \infty} c_k (w - z_0)^k = 0$ and so this sequence of terms is bounded (by Exercise 7.6), say

$$\left| c_k (w - z_0)^k \right| = |c_k| \, r^k \leq M.$$

Now if $|z - z_0| < |w - z_0|$, then

$$\sum_{k \geq 0} \left|c_k(z-z_0)^k\right| = \sum_{k \geq 0} |c_k| \, r^k \left(\frac{|z-z_0|}{r}\right)^k \leq M \sum_{k \geq 0} \left(\frac{|z-z_0|}{r}\right)^k.$$

The sum on the right-hand side is a convergent geometric sequence, since $|z-z_0| < r$, and so $\sum_{k \geq 0} c_k(z-z_0)^k$ converges absolutely by Corollary 7.12. $\square$

Proof of Theorem 7.31. Consider the set

$$S := \left\{ x \in \mathbb{R}_{\geq 0} : \sum_{k \geq 0} c_k \, x^k \text{ converges} \right\}.$$

(This set is nonempty since $0 \in S$.)

If S is unbounded then $\sum_{k \geq 0} c_k(z-z_0)^k$ converges absolutely and uniformly for $|z-z_0| \leq r$, for any r (and so this gives the $R = \infty$ case of Theorem 7.31): choose $x \in S$ with $x > r$, then Proposition 7.32 says that $\sum_{k \geq 0} c_k \, r^k$ converges absolutely. Since $\left|c_k(z-z_0)^k\right| \leq |c_k| r^k$, we can now use Proposition 7.28.

If S is bounded, let R be its least upper bound. If $R = 0$ then $\sum_{k \geq 0} c_k(z-z_0)^k$ converges only for $z = z_0$, which establishes Theorem 7.31 in this case.

Now assume $R > 0$. If $|z-z_0| < R$ then (because R is a least upper bound for S) there exists $r \in S$ such that

$$|z - z_0| < r \leq R.$$

Thus $\sum_{k \geq 0} c_k(w-z_0)^k$ converges for $w = z_0 + r$, and so $\sum_{k \geq 0} c_k(z-z_0)^k$ converges absolutely by Proposition 7.32. This finishes (a).

If $|z-z_0| \leq r$ for some $r < R$, again we can find $x \in S$ such that $r < x \leq R$. Then $\sum_{k \geq 0} |c_k| \, r^k$ converges by Proposition 7.32, and so $\sum_{k \geq 0} c_k(z-z_0)^k$ converges absolutely and uniformly for $|z-z_0| \leq r$ by Proposition 7.28. This proves (b).

Finally, if $|z-z_0| > R$ then there exists $r \notin S$ such that

$$R \leq r < |z-z_0|.$$

But $\sum_{k \geq 0} c_k \, r^k$ diverges, so (by the contrapositive of Theorem 7.20) $\sum_{k \geq 0} |c_k| \, r^k$ diverges, and so (by the contrapositive of Proposition 7.32) $\sum_{k \geq 0} c_k(z-z_0)^k$ diverges, which finishes (c). $\square$

Corollary 7.33. If $\lim_{k\to\infty} \sqrt[k]{|c_k|}$ exists then the radius of convergence of the series $\sum_{k\geq 0} c_k(z-z_0)^k$ equals

$$R = \begin{cases} \infty & \text{if } \lim_{k\to\infty} \sqrt[k]{|c_k|} = 0, \\ \frac{1}{\lim_{k\to\infty} \sqrt[k]{|c_k|}} & \text{otherwise.} \end{cases}$$

Proof. We treat the case that R is finite and leave the case $R = \infty$ to Exercise 7.31.

Given R as in the statement of the corollary, it suffices (by Theorem 7.31) to show that $\sum_{k\geq 0} c_k(z-z_0)^k$ converges for $|z-z_0| < R$ and diverges for $|z-z_0| > R$.

Suppose $r := |z-z_0| < R$. Since $\lim_{k\to\infty} \sqrt[k]{|c_k|} = \frac{1}{R}$ and $\frac{2}{R+r} > \frac{1}{R}$, there exists N such that $\sqrt[k]{|c_k|} < \frac{2}{R+r}$ for $k \geq N$. For those k we then have

$$\left|c_k(z-z_0)^k\right| = |c_k||z-z_0|^k = \left(\sqrt[k]{|c_k|}\, r\right)^k < \left(\frac{2r}{R+r}\right)^k$$

and so $\sum_{k=N}^\infty c_k(z-z_0)^k$ converges (absolutely) by Proposition 7.28, because $\frac{2r}{R+r} < 1$ and thus $\sum_{k\geq 0} \left(\frac{2r}{R+r}\right)^k$ converges as a geometric series. Thus $\sum_{k\geq 0} c_k(z-z_0)^k$ converges.

Now suppose $r = |z-z_0| > R$. Again because $\lim_{k\to\infty} \sqrt[k]{|c_k|} = \frac{1}{R}$ and now $\frac{2}{R+r} < \frac{1}{R}$, there exists N such that $\sqrt[k]{|c_k|} > \frac{2}{R+r}$ for $k \geq N$. For those k,

$$\left|c_k(z-z_0)^k\right| = \left(\sqrt[k]{|c_k|}\, r\right)^k > \left(\frac{2r}{R+r}\right)^k > 1,$$

and so the sequence $c_k(z-z_0)^k$ cannot converge to 0. Subsequently (by Corollary 7.14), $\sum_{k\geq 0} c_k(z-z_0)^k$ diverges. □

You might remember this corollary from calculus, where it goes by the name *root test*. Its twin sister, the *ratio test*, is the subject of Exercise 7.32.

Example 7.34. For the power series $\sum_{k\geq 0} k\, z^k$ we compute

$$\lim_{k\to\infty} \sqrt[k]{|c_k|} = \lim_{k\to\infty} \sqrt[k]{k} = \lim_{k\to\infty} e^{\frac{1}{k}\ln(k)} = e^{\lim_{k\to\infty} \frac{\ln(k)}{k}} = e^0 = 1,$$

and Corollary 7.33 gives the radius of convergence 1. (Alternatively, we can argue by differentiating the geometric series.) □

Example 7.35. Consider the power series $\sum_{k \geq 0} \frac{1}{k!} z^k$. Since

$$\lim_{k \to \infty} \left| \frac{c_{k+1}}{c_k} \right| = \lim_{k \to \infty} \frac{k!}{(k+1)!} = \lim_{k \to \infty} \frac{1}{k+1} = 0,$$

the ratio test (Exercise 7.32) implies that the radius of convergence of $\sum_{k \geq 0} \frac{1}{k!} z^k$ is ∞, and so the power series converges absolutely in $\mathbb{C}$.[5] □

By way of Proposition 7.25, Theorem 7.31 almost immediately implies the following.

Corollary 7.36. Suppose the power series $\sum_{k \geq 0} c_k (z - z_0)^k$ has radius of convergence $R > 0$. Then the series represents a function that is continuous on $D[z_0, R]$.

Proof. Given any point $w \in D[z_0, R]$, we can find $r < R$ such that $w \in D[z_0, r]$ (e.g., if $R \neq \infty$ then $r = \frac{|w - z_0| + R}{2}$ will do the trick). Theorem 7.31 says that $\sum_{k \geq 0} c_k (z - z_0)^k$ converges uniformly in $D[z_0, r]$, and so Proposition 7.25 implies that the power series is continuous in $D[z_0, r]$, and so particularly at w. □

Finally, mixing Proposition 7.27 with Theorem 7.31 gives:

Corollary 7.37. Suppose the power series $\sum_{k \geq 0} c_k (z - z_0)^k$ has radius of convergence $R > 0$ and γ is a piecewise smooth path in $D[z_0, R]$. Then

$$\int_\gamma \sum_{k \geq 0} c_k (z - z_0)^k \, dz = \sum_{k \geq 0} c_k \int_\gamma (z - z_0)^k \, dz.$$

In particular, if γ is closed then $\int_\gamma \sum_{k \geq 0} c_k (z - z_0)^k \, dz = 0$.

Proof. Let $r := \max_{z \in \gamma} |\gamma(z) - z_0|$ (whose existence is guaranteed by Theorem A.1). Then $\gamma \subset \overline{D}[z_0, r]$ and $r < R$. Theorem 7.31 says that $\sum_{k \geq 0} c_k (z - z_0)^k$ converges uniformly in $\overline{D}[z_0, r]$, and so Proposition 7.27 allows us to switch integral and summation.

The last statement follows now with Exercise 4.15. □

These corollaries will become extremely useful in the next chapter.

---

[5] In the next chapter, we will see that this power series represents the exponential function.

Exercises

7.1. For each of the sequences, prove convergence or divergence. If the sequence converges, find the limit.

(a) $a_n = e^{\frac{\pi i n}{4}}$ (c) $a_n = \cos(n)$ (e) $a_n = \sin(\frac{1}{n})$

(b) $a_n = \frac{(-1)^n}{n}$ (d) $a_n = 2 - \frac{i n^2}{2n^2+1}$

7.2. Determine whether each of the following series converges or diverges.

(a) $\sum_{n\geq 1} \left(\frac{1+i}{\sqrt{3}}\right)^n$ (c) $\sum_{n\geq 1} \left(\frac{1+2i}{\sqrt{5}}\right)^n$

(b) $\sum_{n\geq 1} n \left(\frac{1}{i}\right)^n$ (d) $\sum_{n\geq 1} \frac{1}{n^3 + i^n}$

7.3. Compute $\sum_{n\geq 1} \frac{1}{n^2 + 2n}$.

7.4. Prove Proposition 7.3.

7.5. Prove the following:

(a) $\lim_{n\to\infty} a_n = a \implies \lim_{n\to\infty} |a_n| = |a|$.

(b) $\lim_{n\to\infty} a_n = 0 \iff \lim_{n\to\infty} |a_n| = 0$.

7.6. Show that a convergent sequence is bounded, i.e.: if $\lim_{n\to\infty} a_n$ exists, then there is an M such that $|a_n| \leq M$ for all $n \geq 1$.

7.7. Show that the limit of a convergent sequence is unique.

7.8. Let (a_n) be a sequence. A point a is an *accumulation point* of the sequence if for every $\varepsilon > 0$ and every $N \in \mathbb{Z}_{>0}$ there exists some $n > N$ such that $|a_n - a| < \varepsilon$. Prove that if a sequence has more than one accumulation point then the sequence diverges.

7.9.

(a) Show that $\frac{1}{k!} \leq \frac{3}{k(k+1)}$ for any positive integer k.

(b) Conclude with Example 7.9 that for any positive integer n,
$$1 + \frac{1}{2} + \frac{1}{6} + \cdots + \frac{1}{n!} \leq 3.$$

7.10. Derive the Archimedean Property from the Monotone Sequence Property.

7.11. Prove Proposition 7.7.

7.12. Prove that (c_n) converges if and only if $(\operatorname{Re} c_n)$ and $(\operatorname{Im} c_n)$ converge.

7.13. Prove that $\mathbb{Z}$ is complete and that $\mathbb{Q}$ is not complete.

7.14. Prove that, if $a_n \leq b_n \leq c_n$ for all n and $\lim_{n\to\infty} a_n = \lim_{n\to\infty} c_n = L$, then $\lim_{n\to\infty} b_n = L$. This is called the *Squeeze Theorem*, and is useful in testing a sequence for convergence.

7.15. Find the least upper bound of the set $\left\{\operatorname{Re}\left(e^{2\pi i t}\right) : t \in \mathbb{Q} \setminus \mathbb{Z}\right\}$.

7.16.

(a) Suppose that the sequence (c_n) converges to zero. Show that $\sum_{n\geq 0} c_n$ converges if and only if $\sum_{k\geq 0}(c_{2k} + c_{2k+1})$ converges. Moreover, if the two series converge then they have the same limit.

(b) Give an example where (c_n) does not converge to 0 and one of the series in (a) diverges while the other converges.

7.17. Prove that the series $\sum_{k\geq 1} b_k$ converges if and only if $\lim_{n\to\infty} \sum_{k\geq n} b_k = 0$.

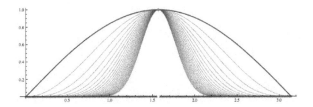

Figure 7.3: The functions $f_n(x) := \sin^n(x)$ in Exercise 7.21.

7.18.

(a) Show that $\displaystyle\sum_{k\geq 1} \frac{k}{k^2+1}$ diverges.

(b) Show that $\displaystyle\sum_{k\geq 1} \frac{k}{k^3+1}$ converges.

7.19.

(a) Suppose $G \subseteq \mathbb{C}$ and $f_n : G \to \mathbb{C}$ for $n \geq 1$. Suppose (a_n) is a sequence in $\mathbb{R}$ with $\lim_{n\to\infty} a_n = 0$ and, for each $n \geq 1$,

$$|f_n(z)| \leq a_n \qquad \text{for all } z \in G.$$

Show that (f_n) converges uniformly to the zero function in G.

(b) Re-prove the statement of Example 7.24 using part (a).

7.20.

(a) Suppose $G \subseteq \mathbb{C}$, $f_n : G \to \mathbb{C}$ for $n \geq 1$, and (f_n) converges uniformly to the zero function in G. Show that, if (z_n) is any sequence in G, then

$$\lim_{n\to\infty} f_n(z_n) = 0.$$

(b) Apply (a) to the function sequence given in Example 7.23, together with the sequence $(z_n = e^{-\frac{1}{n}})$, to prove that the convergence given in Example 7.23 is not uniform.

7.21. Consider $f_n : [0, \pi] \to \mathbb{R}$ given by $f_n(x) = \sin^n(x)$, for $n \geq 1$. Prove that (f_n) converges pointwise to $f : [0, \pi] \to \mathbb{R}$ given by

$$f(x) = \begin{cases} 1 & \text{if } x = \frac{\pi}{2}, \\ 0 & \text{if } x \neq \frac{\pi}{2}, \end{cases}$$

yet this convergence is not uniform. (See Figure 7.3.)

7.22. Where do the following sequences converge pointwise? Do they converge uniformly on this domain?

(a) $(n z^n)$ (b) $\left(\frac{z^n}{n}\right)$ (c) $\left(\frac{1}{1+nz}\right)$ where $\mathrm{Re}(z) \geq 0$

7.23. Let $f_n(x) = n^2 x \, e^{-nx}$.

(a) Show that $\lim_{n \to \infty} f_n(x) = 0$ for all $x \geq 0$. (*Hint*: Treat $x = 0$ as a special case; for $x > 0$ you can use L'Hôspital's rule (Theorem A.11) — but remember that n is the variable, not x.)

(b) Find $\lim_{n \to \infty} \int_0^1 f_n(x) \, dx$. (*Hint*: The answer is *not* 0.)

(c) Why doesn't your answer to part (b) violate Proposition 7.27?

7.24. The product of two power series centered at z_0 is another power series centered at z_0. Derive a formula for its coefficients in terms of the coefficients of the original two power series.

7.25. Find a power series (and determine its radius of convergence) for the following functions. where $z_0 = 0$

(a) $\dfrac{1}{1+4z}$ (b) $\dfrac{1}{3 - \frac{z}{2}}$ (c) $\dfrac{z^2}{(4-z)^2}$

7.26. Find a power series representation about the origin of each of the following functions.

(a) $\cos z$ (b) $\cos(z^2)$ (c) $z^2 \sin z$ (d) $(\sin z)^2$

REGIONS OF CONVERGENCE

7.27.

(a) Suppose that the sequence (c_k) is bounded. Show that the radius of convergence of $\sum_{k \geq 0} c_k(z - z_0)^k$ is at least 1.

(b) Suppose that the sequence (c_k) does not converge to 0. Show that the radius of convergence of $\sum_{k \geq 0} c_k(z - z_0)^k$ is at most 1.

7.28. Find the power series centered at 1 and compute its radius of convergence for each of the following functions:

(a) $f(z) = \frac{1}{z}$ (b) $f(z) = \operatorname{Log}(z)$

7.29. Use the Weierstraß M-test to show that each of the following series converges uniformly on the given domain:

(a) $\sum_{k \geq 1} \frac{z^k}{k^2}$ on $\overline{D}[0, 1]$

(b) $\sum_{k \geq 0} \frac{1}{z^k}$ on $\{z \in \mathbb{C} : |z| \geq 2\}$

(c) $\sum_{k \geq 0} \frac{z^k}{z^k + 1}$ on $\overline{D}[0, r]$ where $0 \leq r < 1$

7.30. Fix $z \in \mathbb{C}$ and $r > |z|$. Prove that $\sum_{k \geq 0} \left(\frac{z}{w}\right)^k$ converges uniformly *in the variable w* for $|w| \geq r$.

7.31. Complete our proof of Corollary 7.33 by considering the case $R = \infty$.

7.32. Prove that, if $\lim_{k \to \infty} \left|\frac{c_{k+1}}{c_k}\right|$ exists then the radius of convergence of the series $\sum_{k \geq 0} c_k(z - z_0)^k$ equals

$$R = \begin{cases} \infty & \text{if } \lim_{k \to \infty} \left|\frac{c_{k+1}}{c_k}\right| = 0, \\ \lim_{k \to \infty} \left|\frac{c_k}{c_{k+1}}\right| & \text{otherwise.} \end{cases}$$

7.33. Find the radius of convergence for each of the following series.

(a) $\sum_{k \geq 0} a^{k^2} z^k$ where $a \in \mathbb{C}$

(b) $\sum_{k \geq 0} k^n z^k$ where $n \in \mathbb{Z}$

(c) $\displaystyle\sum_{k\geq 0} z^{k!}$

(d) $\displaystyle\sum_{k\geq 1} \frac{(-1)^k}{k} z^{k(k+1)}$

(e) $\displaystyle\sum_{k\geq 1} \frac{z^k}{k^k}$

(f) $\displaystyle\sum_{k\geq 0} \cos(k)\, z^k$

(g) $\displaystyle\sum_{k\geq 0} 4^k (z-2)^k$

7.34. Find a function representing each of the following series.

(a) $\displaystyle\sum_{k\geq 0} \frac{z^{2k}}{k!}$

(b) $\displaystyle\sum_{k\geq 1} k(z-1)^{k-1}$

(c) $\displaystyle\sum_{k\geq 2} k(k-1) z^k$

7.35. Recall the function $f : D[0,1] \to \mathbb{C}$ defined in Exercise 5.5 through

$$f(z) := \int_{[0,1]} \frac{dw}{1 - wz}.$$

Find a power series for f.

7.36. Define the functions $f_n : \mathbb{R}_{\geq 0} \to \mathbb{R}$ via $f_n(t) = \frac{1}{n} e^{-\frac{t}{n}}$, for $n \geq 1$.

(a) Show that the maximum of $f_n(t)$ is $\frac{1}{n}$.

(b) Show that $f_n(t)$ converges uniformly to the zero function on $\mathbb{R}_{\geq 0}$.

(c) Show that $\int_0^\infty f_n(t)\, dt$ does not converge to 0 as $n \to \infty$.

(d) Why doesn't this contradict Proposition 7.27?

Chapter 8

Taylor and Laurent Series

We think in generalities, but we live in details.
Alfred North Whitehead (1861–1947)

Now that we have developed some machinery for power series, we are ready to connect them to the earlier chapters. Our first big goal in this chapter is to prove that every power series represents a holomorphic function in its disk of convergence and, vice versa, that every holomorphic function can be locally represented by a power series.

Our second goal returns to our motivation to start Chapter 7: we'd still like to compute (7.1),

$$\int_{C[2,3]} \frac{\exp(z)}{\sin(z)}\, dz\,.$$

Looking back at Figure 7.1 suggests that we expand the function $\frac{\exp(z)}{\sin(z)}$ locally into something like power series centered at 0 and π, and with any luck we can then use Proposition 7.27 to integrate. Of course, $\frac{\exp(z)}{\sin(z)}$ has singularities at 0 and π, so there is no hope of computing power series at these points. We will develop an analogue of a power series centered at a singularity.

8.1 Power Series and Holomorphic Functions

Here is the first (and easier) half of the first goal we just announced.

Theorem 8.1. Suppose $f(z) = \sum_{k \geq 0} c_k (z - z_0)^k$ has radius of convergence $R > 0$. Then f is holomorphic in $D[z_0, R]$.

Proof. Corollary 7.36 says that f is continuous in $D[z_0, R]$. Given any closed piecewise smooth path $\gamma \subset D[z_0, R]$, Corollary 7.37 gives $\int_\gamma f = 0$. Now apply Morera's theorem (Corollary 5.6). □

A special case of this result concerns power series with infinite radius of convergence: those represent entire functions.

Now that we know that power series are differentiable in their regions of convergence, we can ask how to find their derivatives. The next result says that we can simply differentiate the series term by term.

Theorem 8.2. Suppose $f(z) = \sum_{k \geq 0} c_k (z - z_0)^k$ has radius of convergence $R > 0$. Then
$$f'(z) = \sum_{k \geq 1} k\, c_k (z - z_0)^{k-1} \quad \text{for any } z \in D[z_0, R],$$
and the radius of convergence of this power series is also R.

Proof. If $z \in D[z_0, R]$ then $|z - z_0| < R$, so we can choose R_1 so that $|z - z_0| < R_1 < R$. Then the circle $\gamma := C[z_0, R_1]$ lies in $D[z_0, R]$ and z is inside γ. Since f is holomorphic in $D[z_0, R]$ we can use Cauchy's Integral Formula for f' (Theorem 5.1), as well as Corollary 7.37:

$$f'(z) = \frac{1}{2\pi i} \int_\gamma \frac{f(w)}{(w-z)^2}\, dw = \frac{1}{2\pi i} \int_\gamma \frac{1}{(w-z)^2} \sum_{k \geq 0} c_k (w - z_0)^k\, dw$$

$$= \sum_{k \geq 0} c_k \frac{1}{2\pi i} \int_\gamma \frac{(w - z_0)^k}{(w-z)^2}\, dw = \sum_{k \geq 0} c_k \frac{d}{dw}(w - z_0)^k \bigg|_{w=z}$$

$$= \sum_{k \geq 1} k\, c_k (z - z_0)^{k-1}.$$

Note that we used Theorem 5.1 *again* in the penultimate step, but now applied to the function $(z - z_0)^k$.

The last statement of the theorem is easy to show: the radius of convergence of $f'(z)$ is at least R (since we have shown that the series for f' converges whenever $|z - z_0| < R$), and it cannot be larger than R by comparison to the series for $f(z)$, since the coefficients for $(z-z_0) f'(z)$ are larger than the corresponding ones for $f(z)$. □

Example 8.3. Let
$$f(z) = \sum_{k \geq 0} \frac{z^k}{k!}.$$

In Example 7.35, we showed that f converges in $\mathbb{C}$. We claim that $f(z) = \exp(z)$, in analogy with the real exponential function. First, by Theorem 8.2,

$$f'(z) = \frac{d}{dz} \sum_{k \geq 0} \frac{z^k}{k!} = \sum_{k \geq 1} \frac{z^{k-1}}{(k-1)!} = \sum_{k \geq 0} \frac{z^k}{k!} = f(z).$$

Thus

$$\frac{d}{dz} \frac{f(z)}{\exp(z)} = \frac{d}{dz} (f(z)\exp(-z)) = f'(z)\exp(-z) - f(z)\exp(-z) = 0,$$

and so, by Theorem 2.17, $\frac{f(z)}{\exp(z)}$ is constant. Evaluating at $z = 0$ gives that this constant is 1, and so $f(z) = \exp(z)$. □

Example 8.4. We can use the power series expansion for $\exp(z)$ to find power series for the trigonometric functions. For instance,

$$\begin{aligned}
\sin z &= \frac{1}{2i} (\exp(iz) - \exp(-iz)) = \frac{1}{2i} \left(\sum_{k \geq 0} \frac{(iz)^k}{k!} - \sum_{k \geq 0} \frac{(-iz)^k}{k!} \right) \\
&= \frac{1}{2i} \sum_{k \geq 0} \frac{1}{k!} \left((iz)^k - (-1)^k (iz)^k \right) = \frac{1}{2i} \sum_{k \geq 0 \text{ odd}} \frac{2(iz)^k}{k!} \\
&= \frac{1}{i} \sum_{j \geq 0} \frac{(iz)^{2j+1}}{(2j+1)!} = \sum_{j \geq 0} \frac{i^{2j} z^{2j+1}}{(2j+1)!} = \sum_{j \geq 0} \frac{(-1)^j}{(2j+1)!} z^{2j+1} \\
&= z - \frac{z^3}{3!} + \frac{z^5}{5!} - \frac{z^7}{7!} + \cdots.
\end{aligned}$$

Note that we are allowed to rearrange the terms of the two added sums because the corresponding series have infinite radii of convergence. □

Naturally, Theorem 8.2 can be repeatedly applied to f', then to f'', and so on. The various derivatives of a power series can also be seen as ingredients of the series itself—this is the statement of the following *Taylor series expansion*.[1]

Corollary 8.5. *Suppose $f(z) = \sum_{k \geq 0} c_k (z - z_0)^k$ has a positive radius of convergence. Then*

$$c_k = \frac{f^{(k)}(z_0)}{k!}.$$

---

[1] Named after Brook Taylor (1685–1731).

Proof. For starters, $f(z_0) = c_0$. Theorem 8.2 gives $f'(z_0) = c_1$. Applying the same theorem to f' gives

$$f''(z) = \sum_{k \geq 2} k(k-1) c_k (z-z_0)^{k-2}$$

and so $f''(z_0) = 2 c_2$. A quick induction game establishes $f'''(z_0) = 6 c_3$, $f''''(z_0) = 24 c_4$, etc. □

Taylor's formula shows that the coefficients of any power series converging to f on some open disk D can be determined from the function f restricted to D. It follows immediately that the coefficients of a power series are unique:

Corollary 8.6. *If $\sum_{k \geq 0} c_k (z-z_0)^k$ and $\sum_{k \geq 0} d_k (z-z_0)^k$ are two power series that both converge to the same function on an open disk centered at z_0, then $c_k = d_k$ for all $k \geq 0$.*

Example 8.7. We'd like to compute a power series expansion for $f(z) = \exp(z)$ centered at $z_0 = \pi$. Since

$$f^{(k)}(z_0) = \exp(z)\Big|_{z=\pi} = e^\pi,$$

Corollary 8.5 suggests that this power series is

$$\sum_{k \geq 0} \frac{e^\pi}{k!} (z-\pi)^k,$$

which converges for all $z \in \mathbb{C}$ (essentially by Example 7.35). □

We now turn to the second cornerstone result of this section, that a holomorphic function can be locally represented by a power series.

Theorem 8.8. *Suppose f is a function holomorphic in $D[z_0, R]$. Then f can be represented as a power series centered at z_0, with a radius of convergence $\geq R$:*

$$f(z) = \sum_{k \geq 0} c_k (z-z_0)^k \quad \text{with} \quad c_k = \frac{1}{2\pi i} \int_\gamma \frac{f(w)}{(w-z_0)^{k+1}} \, dw,$$

where γ is any positively oriented, simple, closed, piecewise smooth path in $D[z_0, R]$ for which z_0 is inside γ.

Proof. Let $g(z) := f(z + z_0)$; so g is a function holomorphic in $D[0, R]$. Given $z \in D[0, R]$, let $r := \frac{|z|+R}{2}$. By Cauchy's Integral Formula (Theorem 4.27),

$$g(z) = \frac{1}{2\pi i} \int_{C[0,r]} \frac{g(w)}{w - z} dw.$$

The factor $\frac{1}{w-z}$ in this integral can be expanded into a geometric series (note that $w \in C[0, r]$ and so $|\frac{z}{w}| < 1$):

$$\frac{1}{w - z} = \frac{1}{w} \frac{1}{1 - \frac{z}{w}} = \frac{1}{w} \sum_{k \geq 0} \left(\frac{z}{w}\right)^k$$

which converges uniformly in the variable $w \in C[0, r]$ by Exercise 7.30. Hence Proposition 7.27 applies:

$$g(z) = \frac{1}{2\pi i} \int_{C[0,r]} \frac{g(w)}{w - z} dw = \frac{1}{2\pi i} \int_{C[0,r]} g(w) \frac{1}{w} \sum_{k \geq 0} \left(\frac{z}{w}\right)^k dw$$

$$= \sum_{k \geq 0} \left(\frac{1}{2\pi i} \int_{C[0,r]} \frac{g(w)}{w^{k+1}} dw \right) z^k.$$

Now, since $f(z) = g(z - z_0)$, we apply a change of variables to obtain

$$f(z) = \sum_{k \geq 0} \left(\frac{1}{2\pi i} \int_{C[z_0,r]} \frac{f(w)}{(w - z_0)^{k+1}} dw \right) (z - z_0)^k.$$

The only differences of this right-hand side to the statement of the theorem are the paths we're integrating over. However, by Cauchy's Theorem 4.18,

$$\int_{C[z_0,r]} \frac{f(w)}{(w - z_0)^{k+1}} dw = \int_\gamma \frac{f(w)}{(w - z_0)^{k+1}} dw. \qquad \square$$

We note a remarkable feature of our proof: namely, if we are given a holomorphic function $f : G \to \mathbb{C}$ and are interested in expanding f into a power series centered at $z_0 \in G$, then we may maximize the radius of convergence R of this power series, in the sense that its region of convergence reaches to the boundary of G. Let's make this precise.

Definition. For a region $G \subseteq \mathbb{C}$ and a point $z_0 \in G$, we define the *distance of z_0 to ∂G*, the boundary of G, as the greatest lower bound of $\{|z - z_0| : z \in \partial G\}$; if this set is empty, we define the distance of z_0 to ∂G to be ∞.

What we have proved above, on the side, is the following.

Corollary 8.9. If $f : G \to \mathbb{C}$ is holomorphic and $z_0 \in G$, then f can be expanded into a power series centered at z_0 whose radius of convergence is at least the distance of z_0 to ∂G.

Example 8.10. Consider $f : \mathbb{C} \setminus \{\pm i\} \to \mathbb{C}$ given by $f(z) := \frac{1}{z^2+1}$ and $z_0 = 0$. Corollary 8.9 says that the power series expansion of f at 0 will have radius of convergence 1. (Actually, it says this radius is *at least* 1, but it cannot be larger since $\pm i$ are singularities of f.) In fact, we can use a geometric series to compute this power series:

$$f(z) = \frac{1}{z^2+1} = \sum_{k \geq 0} \left(-z^2\right)^k = \sum_{k \geq 0} (-1)^k z^{2k},$$

with radius of convergence 1. □

Corollary 8.9 is yet another example of a result that is plainly false when translated into $\mathbb{R}$; see Exercise 8.6.

Comparing the coefficients of the power series obtained in Theorem 8.8 with those in Corollary 8.5, we arrive at the long-promised extension of Theorems 4.27 and 5.1.

Corollary 8.11. Suppose f is holomorphic in the region G and γ is a positively oriented, simple, closed, piecewise smooth path, such that w is inside γ and $\gamma \sim_G 0$. Then

$$f^{(k)}(w) = \frac{k!}{2\pi i} \int_\gamma \frac{f(z)}{(z-w)^{k+1}} \, dz.$$

Corollary 8.11 combined with our often-used Proposition 4.6((d)) gives an inequality which is often called *Cauchy's Estimate*:

Corollary 8.12. Suppose f is holomorphic in $D[w, R]$ and $|f(z)| \leq M$ for all $z \in D[w, R]$. Then

$$\left|f^{(k)}(w)\right| \leq \frac{k! M}{R^k}.$$

Proof. Let $r < R$. By Corollary 8.11 and Proposition 4.6((d)),

$$\begin{aligned} \left| f^{(k)}(w) \right| &= \left| \frac{k!}{2\pi i} \int_{C[w,r]} \frac{f(z)}{(z-w)^{k+1}} dz \right| \\ &\leq \frac{k!}{2\pi} \max_{z \in C[w,r]} \left| \frac{f(z)}{(z-w)^{k+1}} \right| \text{length}(C[w,r]) \\ &\leq \frac{k!}{2\pi} \frac{M}{r^{k+1}} 2\pi r = \frac{k! M}{r^k}. \end{aligned}$$

The statement now follows since r can be chosen arbitrarily close to R. □

A key aspect of this section is worth emphasizing: namely, we have developed an alternative characterization of what it means for a function to be holomorphic. In Chapter 2, we defined a function to be *holomorphic* in a region G if it is differentiable at each point $z_0 \in G$. We now define what it means for a function to be *analytic* in G.

Definition. Let $f : G \to \mathbb{C}$ and $z_0 \in G$. If there exist $R > 0$ and $c_0, c_1, c_2, \ldots \in \mathbb{C}$ such that the power series

$$\sum_{k \geq 0} c_k (z - z_0)^k$$

converges in $D[z_0, R]$ and agrees with $f(z)$ in $D[z_0, R]$, then f is *analytic at* z_0. We call f *analytic in* G if f is analytic at each point in G.

What we have proved in this section can be summed up as follows:

Theorem 8.13. *For any region G, the class of all analytic functions in G coincides with the class of all holomorphic functions in G.*

While the terms *holomorphic* and *analytic* do not always mean the same thing, in the study of complex analysis they do and are frequently used interchangeably.

8.2 Classification of Zeros and the Identity Principle

When we proved the Fundamental Theorem of Algebra (Theorem 5.11; see also Exercise 5.11), we remarked that, if a polynomial $p(z)$ of degree $d > 0$ has a zero at a (that is, $p(a) = 0$), then $p(z)$ has $z - a$ as a factor. That is, we can write $p(z) = (z - a) q(z)$ where $q(z)$ is a polynomial of degree $d - 1$. We can then ask whether $q(z)$ itself has a zero at a and, if so, we can factor out another $(z - a)$.

Continuing in this way, we see that we can factor $p(z)$ as

$$p(z) = (z-a)^m g(z)$$

where m is a positive integer $\leq d$ and $g(z)$ is a polynomial that does not have a zero at a. The integer m is called the *multiplicity* of the zero a of $p(z)$. Almost exactly the same thing happens for holomorphic functions.

Theorem 8.14 (Classification of Zeros)**.** Suppose $f : G \to \mathbb{C}$ is holomorphic and f has a zero at $a \in G$. Then either

(a) f is identically zero on some open disk D centered at a (that is, $f(z) = 0$ for all $z \in D$); or

(b) there exist a positive integer m and a holomorphic function $g : G \to \mathbb{C}$, such that $g(a) \neq 0$ and

$$f(z) = (z-a)^m g(z) \qquad \text{for all } z \in G.$$

In this case the zero a is isolated: there is a disk $D[a, r]$ which contains no other zero of f.

The integer m in the second case is uniquely determined by f and a and is called the *multiplicity* of the zero at a.

Proof. By Theorem 8.8, there exists $R > 0$ such that we can expand [Power series Thm]

$$f(z) = \sum_{k \geq 0} c_k (z-a)^k \qquad \text{for } z \in D[a, R],$$

[$z_0 = a$]

and $c_0 = f(a) = 0$. There are now exactly two possibilities: either

(a) $c_k = 0$ for all $k \geq 0$; or

(b) there is some positive integer m so that $c_k = 0$ for all $k < m$ but $c_m \neq 0$.

The first case gives $f(z) = 0$ for all $z \in D[a, R]$. So now consider the second case. We note that for $z \in D[a, R]$,

$$\begin{aligned} f(z) &= c_m (z-a)^m + c_{m+1}(z-a)^{m+1} + \cdots = (z-a)^m \left(c_m + c_{m+1}(z-a) + \cdots \right) \\ &= (z-a)^m \sum_{k \geq 0} c_{k+m} (z-a)^k. \end{aligned}$$

Thus we can define a function $g : G \to \mathbb{C}$ through

$$g(z) := \begin{cases} \sum_{k \geq 0} c_{k+m}(z-a)^k & \text{if } z \in D[a, R], \\ \dfrac{f(z)}{(z-a)^m} & \text{if } z \in G \setminus \{a\}. \end{cases}$$

(According to our calculations above, the two definitions give the same value when $z \in D[a, R] \setminus \{a\}$.) The function g is holomorphic in $D[a, R]$ by the first definition, and g is holomorphic in $G \setminus \{a\}$ by the second definition. Note that $g(a) = c_m \neq 0$ and, by construction,

$$f(z) = (z-a)^m g(z) \qquad \text{for all } z \in G.$$

Since $g(a) \neq 0$ there is, by continuity, $r > 0$ so that $g(z) \neq 0$ for all $z \in D[a, r]$, so $D[a, r]$ contains no other zero of f. The integer m is unique, since it is defined in terms of the power series expansion of f at a, which is unique by Corollary 8.6. □

Theorem 8.14 gives rise to the following result, which is sometimes called the *identity principle* or the *uniqueness theorem*.

Theorem 8.15. Suppose G is a region, $f : G \to \mathbb{C}$ is holomorphic, and $f(a_n) = 0$ where (a_n) is a sequence of distinct numbers that converges in G. Then f is the zero function on G.

Applying this theorem to the difference of two functions immediately gives the following variant.

Corollary 8.16. Suppose f and g are holomorphic in a region G and $f(a_k) = g(a_k)$ at a sequence that converges to $w \in G$ with $a_k \neq w$ for all k. Then $f(z) = g(z)$ for all z in G.

Proof of Theorem 8.15. Consider the following two subsets of G:

$X := \{a \in G : \text{there exists } r \text{ such that } f(z) = 0 \text{ for all } z \in D[a, r]\}$
$Y := \{a \in G : \text{there exists } r \text{ such that } f(z) \neq 0 \text{ for all } z \in D[a, r] \setminus \{a\}\}.$

If $f(a) \neq 0$ then, by continuity of f, there exists a disk centered at a in which f is nonzero, and so $a \in Y$.

If $f(a) = 0$, then Theorem 8.14 says that either $a \in X$ or a is an isolated zero of f, so $a \in Y$.

We have thus proved that G is the disjoint union of X and Y. Exercise 8.11 proves that X and Y are open, and so (because G is a region) either X or Y has to be empty. The conditions of Theorem 8.15 say that $\lim_{n \to \infty} a_n$ is not in Y, and thus it has to be in X. Thus $G = X$ and the statement of Theorem 8.15 follows. □

The identity principle yields the strengthenings of Theorem 6.11 and Corollary 6.12 promised in Chapter 6. We recall that that we say the function $u : G \to \mathbb{R}$ has a *weak relative maximum* w if there exists a disk $D[w, r] \subseteq G$ such that all $z \in D[w, r]$ satisfy $u(z) \leq u(w)$.

Theorem 8.17 (Maximum-Modulus Theorem). Suppose f is holomorphic and nonconstant in a region G. Then $|f|$ does not attain a weak relative maximum in G.

There are many reformulations of this theorem, such as:

Corollary 8.18. Suppose f is holomorphic in a bounded region G and continuous on its closure. Then
$$\sup_{z \in G} |f(z)| = \max_{z \in \partial G} |f(z)|.$$

Proof. This is trivial if f is constant, so we assume f is non-constant. By the Extreme Value Theorem A.1 there is a point $z_0 \in \overline{G}$ so that $\max_{z \in \overline{G}} |f(z)| = |f(z_0)|$. Clearly $\sup_{z \in G} |f(z)| \leq \max_{z \in \overline{G}} |f(z)|$, and this is easily seen to be an equality using continuity at z_0, since there are points of G arbitrarily close to z_0. Finally, Theorem 8.17 implies $z_0 \notin G$, so z_0 must be in ∂G. □

Theorem 8.17 has other important consequences; we give two here, whose proofs we leave for Exercises 8.12 and 8.13.

Corollary 8.19 (Minimum-Modulus Theorem). Suppose f is holomorphic and nonconstant in a region G. Then $|f|$ does not attain a weak relative minimum at a point a in G unless $f(a) = 0$.

Corollary 8.20. If u is harmonic and nonconstant in a region G, then it does not have a weak relative maximum or minimum in G.

Note that Equation (6.1) in Chapter 6 follows from Corollary 8.20 using essentially the same argument as in the proof of Corollary 8.18.

Proof of Theorem 8.17. Suppose there exist $a \in G$ and $R > 0$ such that $|f(a)| \geq |f(z)|$ for all $z \in D[a, R]$. We will show that then f is constant.

If $f(a) = 0$ then $f(z) = 0$ for all $z \in D[a, R]$, so f is identically zero by Theorem 8.15.

Now assume $f(a) \neq 0$, which allows us to define the holomorphic function $g : G \to \mathbb{C}$ via $g(z) := \frac{f(z)}{f(a)}$. This function satisfies

$$|g(z)| \leq |g(a)| = 1 \qquad \text{for all } z \in D[a, R],$$

Also $g(a) = 1$, so, by continuity of g, we can find $r \leq R$ such that $\mathrm{Re}(g(z)) > 0$ for $z \in D[a, r]$. This allows us, in turn, to define the holomorphic function $h : D[a, r] \to \mathbb{C}$ through $h(z) := \mathrm{Log}(g(z))$, which satisfies

$$h(a) = \mathrm{Log}(g(a)) = \mathrm{Log}(1) = 0$$

and

$$\mathrm{Re}(h(z)) = \mathrm{Re}(\mathrm{Log}(g(z))) = \ln(|g(z)|) \leq \ln(1) = 0.$$

Exercise 8.35 now implies that h must be identically zero in $D[a, r]$. Hence $g(z) = \exp(h(z))$ must be equal to $\exp(0) = 1$ for all $z \in D[a, r]$, and so $f(z) = f(a) g(z)$ must have the constant value $f(a)$ for all $z \in D[a, r]$. Corollary 8.16 then implies that f is constant in G. □

8.3 Laurent Series

Theorem 8.8 gives a powerful way of describing holomorphic functions. It is, however, not as general as it could be. It is natural, for example, to think about representing $\exp(\frac{1}{z})$ as

$$\exp\left(\frac{1}{z}\right) = \sum_{k \geq 0} \frac{1}{k!} \left(\frac{1}{z}\right)^k = \sum_{k \geq 0} \frac{1}{k!} z^{-k},$$

a "power series" with negative exponents. To make sense of expressions like the above, we introduce the concept of a *double series*

$$\sum_{k \in \mathbb{Z}} a_k := \sum_{k \geq 0} a_k + \sum_{k \geq 1} a_{-k}.$$

Here $a_k \in \mathbb{C}$ are terms indexed by the integers. The double series above *converges* if and only if the two series on the right-hand side do. Absolute and uniform convergence are defined analogously. Equipped with this, we can now introduce the following central concept.

Definition. A *Laurent[2] series centered at* z_0 is a double series of the form

$$\sum_{k \in \mathbb{Z}} c_k (z - z_0)^k.$$

Example 8.21. The series that started this section is the Laurent series of $\exp(\frac{1}{z})$ centered at 0. □

Example 8.22. Any power series is a Laurent series (with $c_k = 0$ for $k < 0$). □

We should pause for a minute and ask for which z a general Laurent series can possibly converge. By definition

$$\sum_{k \in \mathbb{Z}} c_k (z - z_0)^k = \sum_{k \geq 0} c_k (z - z_0)^k + \sum_{k \geq 1} c_{-k} (z - z_0)^{-k}.$$

The first series on the right-hand side is a power series with some radius of convergence R_2, that is, with Theorem 7.31, it converges in $\{z \in \mathbb{C} : |z - z_0| < R_2\}$, and the convergence is uniform in $\{z \in \mathbb{C} : |z - z_0| \leq r_2\}$, for any fixed $r_2 < R_2$. For the second series, we invite you (in Exercise 8.30) to revise our proof of Theorem 7.31 to show that this series converges for

$$\frac{1}{|z - z_0|} < \frac{1}{R_1}$$

for some R_1, and that the convergence is uniform in $\{z \in \mathbb{C} : |z - z_0| \geq r_1\}$, for any fixed $r_1 > R_1$. Thus the Laurent series converges in the annulus

$$A := \{z \in \mathbb{C} : R_1 < |z - z_0| < R_2\}$$

---

[2] After Pierre Alphonse Laurent (1813–1854).

(assuming this set is not empty, i.e., $R_1 < R_2$), and the convergence is uniform on any set of the form

$$\{z \in \mathbb{C} : r_1 \leq |z - z_0| \leq r_2\} \quad \text{for } R_1 < r_1 < r_2 < R_2.$$

Example 8.23. We'd like to compute the start of a Laurent series for $\frac{1}{\sin(z)}$ centered at $z_0 = 0$. We start by considering the function $g : D[0, \pi] \to \mathbb{C}$ defined by

$$g(z) := \begin{cases} \frac{1}{\sin(z)} - \frac{1}{z} & \text{if } z \neq 0, \\ 0 & \text{if } z = 0. \end{cases}$$

A quick application of L'Hôspital's Rule (A.11) shows that g is continuous (see Exercise 8.31). Even better, another round of L'Hôspital's Rule proves that

$$\lim_{z \to 0} \frac{\frac{1}{\sin(z)} - \frac{1}{z}}{z} = \frac{1}{6}.$$

But this means that

$$g'(z) = \begin{cases} -\frac{\cos(z)}{\sin^2(z)} + \frac{1}{z^2} & \text{if } z \neq 0, \\ \frac{1}{6} & \text{if } z = 0, \end{cases}$$

in particular, g is holomorphic in $D[0, \pi]$.[3] By Theorem 8.8, g has a power series expansion at 0, which we may compute using Corollary 8.5. It starts with

$$g(z) = \frac{1}{6} z + \frac{7}{360} z^3 + \frac{31}{15120} z^5 + \cdots$$

and it converges, by Corollary 8.9, for $|z| < \pi$. But this gives our sought-after Laurent series

$$\frac{1}{\sin(z)} = z^{-1} + \frac{1}{6} z + \frac{7}{360} z^3 + \frac{31}{15120} z^5 + \cdots$$

which converges for $0 < |z| < \pi$. □

---

[3] This is a (simple) example of *analytic continuation*: the function g is holomorphic in $D[0, \pi]$ and agrees with $\frac{1}{\sin(z)} - \frac{1}{z}$ in $D[0, \pi] \setminus \{0\}$, the domain in which the latter function is holomorphic. When we said, in the footnote on p. 131, that the Riemann zeta function $\zeta(z) = \sum_{k \geq 1} \frac{1}{k^z}$ can be extended to a function that is holomorphic on $\mathbb{C} \setminus \{1\}$, we were also talking about analytic continuation.

Theorem 8.1 implies that a Laurent series represents a function that is holomorphic in its annulus of convergence. The fact that we can conversely represent any function holomorphic in such an annulus by a Laurent series is the substance of the next result.

Theorem 8.24. Suppose f is a function that is holomorphic in

$$A := \{z \in \mathbb{C} : R_1 < |z - z_0| < R_2\}.$$

Then f can be represented in A as a Laurent series centered at z_0,

$$f(z) = \sum_{k \in \mathbb{Z}} c_k (z - z_0)^k \quad \text{with} \quad c_k = \frac{1}{2\pi i} \int_{C[z_0, r]} \frac{f(w)}{(w - z_0)^{k+1}} dw,$$

where $R_1 < r < R_2$.

By Cauchy's Theorem 4.18 we can replace the circle $C[z_0, r]$ in the formula for the Laurent coefficients by any path $\gamma \sim_A C[z_0, r]$.

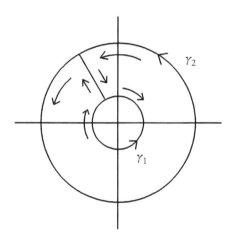

Figure 8.1: The path γ in our proof of Theorem 8.24.

Proof. Let $g(z) = f(z + z_0)$; so g is holomorphic in $\{z \in \mathbb{C} : R_1 < |z| < R_2\}$. Fix $R_1 < r_1 < |z| < r_2 < R_2$, and let γ be the path in Figure 8.1, where $\gamma_1 := C[0, r_1]$

and $\gamma_2 := C[0, r_2]$. By Cauchy's Integral Formula (Theorem 4.27),

$$g(z) = \frac{1}{2\pi i} \int_\gamma \frac{g(w)}{w-z} dw = \frac{1}{2\pi i} \int_{\gamma_2} \frac{g(w)}{w-z} dw - \frac{1}{2\pi i} \int_{\gamma_1} \frac{g(w)}{w-z} dw. \quad (8.1)$$

For the integral over γ_2 we play exactly the same game as in our proof of Theorem 8.8. The factor $\frac{1}{w-z}$ in this integral can be expanded into a geometric series (note that $w \in \gamma_2$ and so $|\frac{z}{w}| < 1$)

$$\frac{1}{w-z} = \frac{1}{w} \frac{1}{1-\frac{z}{w}} = \frac{1}{w} \sum_{k \geq 0} \left(\frac{z}{w}\right)^k,$$

which converges uniformly in the variable $w \in \gamma_2$ by Exercise 7.30. Hence Proposition 7.27 applies:

$$\int_{\gamma_2} \frac{g(w)}{w-z} dw = \int_{\gamma_2} g(w) \frac{1}{w} \sum_{k \geq 0} \left(\frac{z}{w}\right)^k dw = \sum_{k \geq 0} \left(\int_{\gamma_2} \frac{g(w)}{w^{k+1}} dw\right) z^k.$$

The integral over γ_1 is computed in a similar fashion; now we expand the factor $\frac{1}{w-z}$ into the following geometric series (note that $w \in \gamma_1$ and so $|\frac{w}{z}| < 1$)

$$\frac{1}{w-z} = -\frac{1}{z} \frac{1}{1-\frac{w}{z}} = -\frac{1}{z} \sum_{k \geq 0} \left(\frac{w}{z}\right)^k,$$

which converges uniformly in the variable $w \in \gamma_1$. Again Proposition 7.27 applies:

$$\int_{\gamma_1} \frac{g(w)}{w-z} dw = -\int_{\gamma_1} g(w) \frac{1}{z} \sum_{k \geq 0} \left(\frac{w}{z}\right)^k dw = -\sum_{k \geq 0} \left(\int_{\gamma_1} g(w) w^k dw\right) z^{-k-1}$$

$$= -\sum_{k \leq -1} \left(\int_{\gamma_1} \frac{g(w)}{w^{k+1}} dw\right) z^k.$$

Putting everything back into (8.1) gives

$$g(z) = \frac{1}{2\pi i} \left(\sum_{k \geq 0} \left(\int_{\gamma_2} \frac{g(w)}{w^{k+1}} dw\right) z^k + \sum_{k \leq -1} \left(\int_{\gamma_1} \frac{g(w)}{w^{k+1}} dw\right) z^k \right).$$

By Cauchy's Theorem 4.18, we can now change both γ_1 and γ_2 to $C[0, r]$, as long as $R_1 < r < R_2$, which finally gives

$$g(z) = \frac{1}{2\pi i} \sum_{k \in \mathbb{Z}} \left(\int_{C[0,r]} \frac{g(w)}{w^{k+1}} dw \right) z^k .$$

The statement follows now with $f(z) = g(z - z_0)$ and a change of variables in the integral. □

This theorem, naturally, has several corollaries that have analogues in the world of Taylor series. Here are two samples:

Corollary 8.25. If $\sum_{k \in \mathbb{Z}} c_k (z - z_0)^k$ and $\sum_{k \in \mathbb{Z}} d_k (z - z_0)^k$ are two Laurent series that both converge, for $R_1 < |z - z_0| < R_2$, to the same function, then $c_k = d_k$ for all $k \in \mathbb{Z}$.

Corollary 8.26. If G is a region, $z_0 \in G$, and f is holomorphic in $G \setminus \{z_0\}$, then f can be expanded into a Laurent series centered at z_0 that converges for $0 < |z - z_0| < R$ where R is at least the distance of z_0 to ∂G.

Finally, we come to the analogue of Corollary 7.37 for Laurent series. We could revisit its proof, but the statement that would follow is actually the special case $k = -1$ of Theorem 8.24, read from right to left:

Corollary 8.27. Suppose f is a function that is holomorphic in

$$A := \{z \in \mathbb{C} : R_1 < |z - z_0| < R_2\}$$

with Laurent series

$$f(z) = \sum_{k \in \mathbb{Z}} c_k (z - z_0)^k .$$

If γ is any simple, closed, piecewise smooth path in A, such that z_0 is inside γ, then

$$\int_\gamma f(z) \, dz = 2\pi i \, c_{-1} .$$

This result is profound: it says that we can integrate (at least over closed curves) by computing Laurent series—in fact, we "only" need to compute *one coefficient* of a Laurent series. We will have more to say about this in the next chapter; for now, we

LAURENT SERIES

give just one application, which might have been bugging you since the beginning of Chapter 7.

Example 8.28. We will (finally!) compute (7.1), the integral $\int_{C[2,3]} \frac{\exp(z)}{\sin(z)} \, dz$. Our plan is to split up the integration path $C[2,3]$ as in Figure 7.1, which gives, say,

$$\int_{C[2,3]} \frac{\exp(z)}{\sin(z)} \, dz = \int_{C[0,1]} \frac{\exp(z)}{\sin(z)} \, dz + \int_{C[\pi,1]} \frac{\exp(z)}{\sin(z)} \, dz.$$

To compute the two integrals on the right-hand side, we can use Corollary 8.27, for which we need the Laurent expansions of $\frac{\exp(z)}{\sin(z)}$ centered at 0 and π.

By Examples 8.3 and 8.23,

$$\frac{\exp(z)}{\sin(z)} = \left(1 + z + \frac{1}{2}z^2 + \frac{1}{6}z^3 + \cdots\right)\left(z^{-1} + \frac{1}{6}z + \frac{7}{360}z^3 + \frac{31}{15120}z^5 + \cdots\right)$$

$$= z^{-1} + 1 + \frac{2}{3}z + \cdots$$

and Corollary 8.27 gives $\int_{C[0,1]} \frac{\exp(z)}{\sin(z)} \, dz = 2\pi i$.

For the integral around π, we use the fact that $\sin(z) = \sin(\pi - z)$, and so we can compute the Laurent expansion of $\frac{1}{\sin(z)}$ at π also via Example 8.23:

$$\frac{1}{\sin(z)} = -\frac{1}{\sin(z-\pi)} = -(z-\pi)^{-1} - \frac{1}{6}(z-\pi) - \frac{7}{360}(z-\pi)^3 - \cdots$$

Adding Example 8.7 to the mix yields

$$\frac{\exp(z)}{\sin(z)} = \left(e^\pi + e^\pi(z-\pi) + \frac{e^\pi}{2}(z-\pi)^2 + \cdots\right)\left(-(z-\pi)^{-1} - \frac{1}{6}(z-\pi) - \cdots\right)$$

$$= -e^\pi(z-\pi)^{-1} - e^\pi - \frac{2}{3}e^\pi(z-\pi) + \cdots$$

and now Corollary 8.27 gives $\int_{C[\pi,1]} \frac{\exp(z)}{\sin(z)} \, dz = -2\pi i \, e^\pi$. Putting it all together yields the integral we've been after for two chapters:

$$\int_{C[2,3]} \frac{\exp(z)}{\sin(z)} \, dz = 2\pi i \, (1 - e^\pi). \qquad \square$$

Exercises

8.1. For each of the following series, determine where the series converges absolutely and where it converges uniformly:

(a) $\sum_{k\geq 0} \frac{1}{(2k+1)!} z^{2k+1}$

(b) $\sum_{k\geq 0} \left(\frac{1}{z-3}\right)^k$

8.2. What functions are represented by the series in the previous exercise?

8.3. Find the power series centered at π for $\sin(z)$.

8.4. Re-prove Proposition 3.16 using the power series of $\exp(z)$ centered at 0.

8.5. Find the terms through third order and the radius of convergence of the power series for each of the following functions, centered at z_0. (Do not find the general form for the coefficients.)

(a) $f(z) = \dfrac{1}{1+z^2}$, $z_0 = 1$

(b) $f(z) = \dfrac{1}{\exp(z)+1}$, $z_0 = 0$

(c) $f(z) = (1+z)^{\frac{1}{2}}$, $z_0 = 0$

(d) $f(z) = \exp(z^2)$, $z_0 = i$

8.6. Consider $f : \mathbb{R} \to \mathbb{R}$ given by $f(x) := \frac{1}{x^2+1}$, the real version of our function in Example 8.10, to show that Corollary 8.9 has no analogue in $\mathbb{R}$.[4]

8.7. Prove the following generalization of Theorem 8.1: Suppose (f_n) is a sequence of functions that are holomorphic in a region G, and (f_n) converges uniformly to f on G. Then f is holomorphic in G. (This result is called the *Weierstraß convergence theorem*.)

8.8. Use the previous exercise and Corollary 8.12 to prove the following: Suppose (f_n) is a sequence of functions that are holomorphic in a region G and that (f_n) converges uniformly to f on G. Then for any $k \in \mathbb{N}$, the sequence of kth derivatives $\left(f_n^{(k)}\right)$ converges (pointwise) to $f^{(k)}$.

---

[4] Incidentally, the same example shows, once more, that Liouville's theorem (Corollary 5.13) has no analogue in $\mathbb{R}$.

8.9. Suppose $|c_k| \geq 2^k$ for all k. What can you say about the radius of convergence of $\sum_{k \geq 0} c_k z^k$?

8.10. Suppose the radius of convergence of $\sum_{k \geq 0} c_k z^k$ is R. What is the radius of convergence of each of the following?

(a) $\displaystyle\sum_{k \geq 0} c_k z^{2k}$

(b) $\displaystyle\sum_{k \geq 0} 3^k c_k z^k$

(c) $\displaystyle\sum_{k \geq 0} c_k z^{k+5}$

(d) $\displaystyle\sum_{k \geq 0} k^2 c_k z^k$

8.11. Suppose G is a region and $f : G \to \mathbb{C}$ is holomorphic. Prove that the sets

$$X = \{a \in G : \text{there exists } r \text{ such that } f(z) = 0 \text{ for all } z \in D[a, r]\}$$
$$Y = \{a \in G : \text{there exists } r \text{ such that } f(z) \neq 0 \text{ for all } z \in D[a, r] \setminus \{a\}\}.$$

in our proof of Theorem 8.15 are open.

8.12. Prove the Minimum-Modulus Theorem (Corollary 8.19): Suppose f is holomorphic and nonconstant in a region G. Then $|f|$ does not attain a weak relative minimum at a point a in G unless $f(a) = 0$.

8.13. Prove Corollary 8.20: Assume that u is harmonic in a region G and has a weak local maximum at $a \in G$.

(a) If G is simply connected then apply Theorem 8.17 to $\exp(u(z) + iv(z))$, where v is a harmonic conjugate of u. Conclude that u is constant on G.

(b) If G is not simply connected, then the above argument applies to u on *any* disk $D[a, R] \subset G$. Conclude that the partials u_x and u_y are zero on G, and adapt the argument of Theorem 2.17 to show that u is constant.

8.14. Let $f : \mathbb{C} \to \mathbb{C}$ be given by $f(z) = z^2 - 2$. Find the maximum and minimum of $|f(z)|$ on the closed unit disk.

8.15. Give another proof of the Fundamental Theorem of Algebra (Theorem 5.11), using the Minimum-Modulus Theorem (Corollary 8.19). (*Hint*: Use Proposition 5.10 to show that a polynomial does not achieve its minimum modulus on a large circle; then use the Minimum-Modulus Theorem to deduce that the polynomial has a zero.)

8.16. Give another proof of (a variant of) the Maximum-Modulus Theorem 8.17 via Corollary 8.11, as follows: Suppose f is holomorphic in a region containing $\overline{D}[a, r]$, and $|f(z)| \leq M$ for $z \in C[a, r]$. Given a point $z_0 \in D[a, r]$, show (e.g., by Corollary 8.11) that there is a constant $c \in \mathbb{C}$ such that

$$\left|f(z_0)^k\right| \leq c M^k.$$

Conclude that $|f(z_0)| \leq M$.

8.17. Find a Laurent series for

$$\frac{1}{(z-1)(z+1)}$$

centered at $z = 1$ and specify the region in which it converges.

8.18. Find a Laurent series for

$$\frac{1}{z(z-2)^2}$$

centered at $z = 2$ and specify the region in which it converges.

8.19. Find a Laurent series for $\dfrac{z-2}{z+1}$ centered at $z = -1$ and the region in which it converges.

8.20. Find the terms $c_n z^n$ in the Laurent series for $\dfrac{1}{\sin^2(z)}$ centered at $z = 0$, for $-4 \leq n \leq 4$.

8.21. Find the first four nonzero terms in the power series expansion of $\tan(z)$ centered at the origin. What is the radius of convergence?

8.22.

(a) Find the power series representation for $\exp(az)$ centered at 0, where $a \in \mathbb{C}$ is any constant.

(b) Show that
$$\exp(z)\cos(z) = \frac{1}{2}(\exp((1+i)z) + \exp((1-i)z)).$$

(c) Find the power series expansion for $\exp(z)\cos(z)$ centered at 0.

8.23. Show that
$$\frac{z-1}{z-2} = \sum_{k \geq 0} \frac{1}{(z-1)^k}$$
for $|z-1| > 1$.

8.24. Prove: If f is entire and $\mathrm{Im}(f)$ is constant on the closed unit disk then f is constant.

8.25.

(a) Find the Laurent series for $\frac{\cos z}{z^2}$ centered at $z = 0$.

(b) Prove that $f : \mathbb{C} \to \mathbb{C}$ is entire, where
$$f(z) = \begin{cases} \frac{\cos z - 1}{z^2} & \text{if } z \neq 0, \\ -\frac{1}{2} & \text{if } z = 0. \end{cases}$$

8.26. Find the Laurent series for $\sec(z)$ centered at the origin.

8.27. Suppose that f is holomorphic at z_0, $f(z_0) = 0$, and $f'(z_0) \neq 0$. Show that f has a zero of multiplicity 1 at z_0.

8.28. Find the multiplicities of the zeros of each of the following functions:

(a) $f(z) = \exp(z) - 1$, $z_0 = 2k\pi i$, where k is any integer.

(b) $f(z) = \sin(z) - \tan(z)$, $z_0 = 0$.

(c) $f(z) = \cos(z) - 1 + \frac{1}{2}\sin^2(z)$, $z_0 = 0$.

8.29. Find the zeros of the following functions and determine their multiplicities:

(a) $(1+z^2)^4$

(b) $\sin^2(z)$

(c) $1+\exp(z)$

(d) $z^3\cos(z)$

8.30. Prove that the series of the negative-index terms of a Laurent series

$$\sum_{k\geq 1} c_{-k}(z-z_0)^{-k}$$

converges for

$$\frac{1}{|z-z_0|} < \frac{1}{R_1}$$

for some R_1, and that the convergence is uniform in $\{z \in \mathbb{C} : |z-z_0| \geq r_1\}$, for any fixed $r_1 > R_1$.

8.31. Show that

$$\lim_{z\to 0}\left(\frac{1}{\sin(z)} - \frac{1}{z}\right) = 0$$

and

$$\lim_{z\to 0} \frac{\frac{1}{\sin(z)} - \frac{1}{z}}{z} = \frac{1}{6}.$$

(These are the limits we referred to in Example 8.23.)

8.32. Find the three Laurent series of

$$f(z) = \frac{3}{(1-z)(z+2)},$$

centered at 0, defined on the three regions $|z| < 1$, $1 < |z| < 2$, and $2 < |z|$, respectively. (*Hint*: Use a partial fraction decomposition.)

8.33. Suppose that $f(z)$ has exactly one zero, at a, inside the circle γ, and that it has multiplicity 1. Show that

$$a = \frac{1}{2\pi i}\int_\gamma \frac{z f'(z)}{f(z)}\,dz.$$

8.34. Recall that a function $f : G \to \mathbb{C}$ is *even* if $f(-z) = f(z)$ for all $z \in G$, and f is *odd* if $f(-z) = -f(z)$ for all $z \in G$. Prove that, if f is even (resp., odd), then the Laurent series of f at 0 has only even (resp., odd) powers.

8.35. Suppose f is holomorphic and not identically zero on an open disk D centered at a, and suppose $f(a) = 0$. Use the following outline to show that $\operatorname{Re} f(z) > 0$ for some z in D.

  (a) Why can you write $f(z) = (z-a)^m g(z)$ where $m > 0$, g is holomorphic, and $g(a) \neq 0$?

  (b) Write $g(a)$ in polar coordinates as $c\, e^{i\alpha}$ and define $G(z) = e^{-i\alpha} g(z)$. Why is $\operatorname{Re} G(a) > 0$?

  (c) Why is there a positive constant δ so that $\operatorname{Re} G(z) > 0$ for all $z \in D[a, \delta]$?

  (d) Write $z = a + r e^{i\theta}$ for $0 < r < \delta$. Show that $f(z) = r^m e^{im\theta} e^{i\alpha} G(z)$.

  (e) Find a value of θ so that $f(z)$ has positive real part.

8.36.

  (a) Find a Laurent series for
  $$\frac{1}{(z^2 - 4)(z - 2)}$$
  centered at $z = 2$ and specify the region in which it converges.

  (b) Compute $\displaystyle\int_{C[2,1]} \frac{dz}{(z^2 - 4)(z - 2)}$.

8.37.

  (a) Find the power series of $\exp(z)$ centered at $z = -1$.

  (b) Compute $\displaystyle\int_{C[-2,2]} \frac{\exp(z)}{(z+1)^{34}}\, dz$.

8.38. Compute $\displaystyle\int_\gamma \frac{\exp(z)}{\sin(z)}\, dz$ where γ is a closed curve not passing through integer multiples of π.

Chapter 9

Isolated Singularities and the Residue Theorem

> $\frac{1}{r^2}$ *has a nasty singularity at r = 0, but it did not bother Newton—the moon is far enough.*
> Edward Witten

We return one last time to the starting point of Chapters 7 and 8: the quest for

$$\int_{C[2,3]} \frac{\exp(z)}{\sin(z)} \, dz \, .$$

We computed this integral in Example 8.28 crawling on hands and knees (but we finally computed it!), by considering various Taylor and Laurent expansions of $\exp(z)$ and $\frac{1}{\sin(z)}$. In this chapter, we develop a calculus for similar integral computations.

9.1 Classification of Singularities

What are the differences among the functions $\frac{\exp(z)-1}{z}$, $\frac{1}{z^4}$, and $\exp(\frac{1}{z})$ at $z = 0$? None of them are defined at 0, but each singularity is of a different nature. We will frequently consider functions in this chapter that are holomorphic in a disk except at its center (usually because that's where a singularity lies), and it will be handy to define the *punctured disk* with center z_0 and radius R,

$$\dot{D}[z_0, R] := \{z \in \mathbb{C} : 0 < |z - z_0| < R\} = D[z_0, R] \setminus \{z_0\}.$$

CLASSIFICATION OF SINGULARITIES

We extend this definition naturally with $D[z_0, \infty] := \mathbb{C} \setminus \{z_0\}$. For complex functions there are three types of singularities, which are classified as follows.

Definition. If f is holomorphic in the punctured disk $D[z_0, R]$ for some $R > 0$ but not at $z = z_0$, then z_0 is an *isolated singularity* of f. The singularity z_0 is called

(a) *removable* if there exists a function g holomorphic in $D[z_0, R]$ such that $f = g$ in $\mathcal{D}[z_0, R]$,

(b) a *pole* if $\lim\limits_{z \to z_0} |f(z)| = \infty$,

(c) *essential* if z_0 is neither removable nor a pole.

Example 9.1. Let $f : \mathbb{C} \setminus \{0\} \to \mathbb{C}$ be given by $f(z) = \frac{\exp(z)-1}{z}$. Since

$$\exp(z) - 1 = \sum_{k \geq 1} \frac{1}{k!} z^k,$$

the function $g : \mathbb{C} \to \mathbb{C}$ defined by

$$g(z) := \sum_{k \geq 0} \frac{1}{(k+1)!} z^k,$$

which is entire (because this power series converges in $\mathbb{C}$), agrees with f in $\mathbb{C} \setminus \{0\}$. Thus f has a removable singularity at 0. □

Example 9.2. In Example 8.23, we showed that $f : \mathbb{C} \setminus \{j\pi : j \in \mathbb{Z}\} \to \mathbb{C}$ given by $f(z) = \frac{1}{\sin(z)} - \frac{1}{z}$ has a removable singularity at 0, because we proved that $g : D[0, \pi] \to \mathbb{C}$ defined by

$$g(z) = \begin{cases} \frac{1}{\sin(z)} - \frac{1}{z} & \text{if } z \neq 0, \\ 0 & \text{if } z = 0. \end{cases}$$

is holomorphic in $D[0, \pi]$ and agrees with f on $\mathcal{D}[0, \pi]$. □

Example 9.3. The function $f : \mathbb{C} \setminus \{0\} \to \mathbb{C}$ given by $f(z) = \frac{1}{z^4}$ has a pole at 0, as

$$\lim_{z \to 0} \left| \frac{1}{z^4} \right| = \infty.$$

□

Example 9.4. The function $f : \mathbb{C}\setminus\{0\} \to \mathbb{C}$ given by $f(z) = \exp(\frac{1}{z})$ has an essential singularity at 0: the two limits

$$\lim_{x \to 0^+} \exp\left(\frac{1}{x}\right) = \infty \quad \text{and} \quad \lim_{x \to 0^-} \exp\left(\frac{1}{x}\right) = 0$$

show that f has neither a removable singularity nor a pole. □

To get a feel for the different types of singularities, we start with the following criteria.

Proposition 9.5. *Suppose z_0 is an isolated singularity of f. Then*

(a) *z_0 is removable if and only if $\lim_{z \to z_0} (z - z_0) f(z) = 0$;*

(b) *z_0 is a pole if and only if it is not removable and $\lim_{z \to z_0} (z - z_0)^{n+1} f(z) = 0$ for some positive integer n.*

Proof. (a) Suppose that z_0 is a removable singularity of f, so there exists a holomorphic function h on $D[z_0, R]$ such that $f(z) = h(z)$ for all $z \in D[z_0, R]$. But then h is continuous at z_0, and so

$$\lim_{z \to z_0} (z - z_0) f(z) = \lim_{z \to z_0} (z - z_0) h(z) = h(z_0) \lim_{z \to z_0} (z - z_0) = 0.$$

Conversely, suppose that $\lim_{z \to z_0} (z - z_0) f(z) = 0$ and f is holomorphic in $D[z_0, R]$. We define the function $g : D[z_0, R] \to \mathbb{C}$ by

$$g(z) := \begin{cases} (z - z_0)^2 f(z) & \text{if } z \neq z_0, \\ 0 & \text{if } z = z_0. \end{cases}$$

Then g is holomorphic in $D[z_0, R]$ and

$$g'(z_0) = \lim_{z \to z_0} \frac{g(z) - g(z_0)}{z - z_0} = \lim_{z \to z_0} \frac{(z - z_0)^2 f(z)}{z - z_0} = \lim_{z \to z_0} (z - z_0) f(z) = 0,$$

so g is holomorphic in $D[z_0, R]$. We can thus expand g into a power series

$$g(z) = \sum_{k \geq 0} c_k (z - z_0)^k$$

CLASSIFICATION OF SINGULARITIES 173

whose first two terms are zero: $c_0 = g(z_0) = 0$ and $c_1 = g'(z_0) = 0$. But then we can write

$$g(z) = (z - z_0)^2 \sum_{k \geq 0} c_{k+2} (z - z_0)^k$$

and so

$$f(z) = \sum_{k \geq 0} c_{k+2} (z - z_0)^k \qquad \text{for all } z \in D[z_0, R].$$

But this power series is holomorphic in $D[z_0, R]$, so z_0 is a removable singularity.

(b) Suppose that z_0 is a pole of f. Since $f(z) \to \infty$ as $z \to z_0$ we may assume that R is small enough that $f(z) \neq 0$ for $z \in D[z_0, R]$. Then $\frac{1}{f}$ is holomorphic in $D[z_0, R]$ and

$$\lim_{z \to z_0} \frac{1}{f(z)} = 0,$$

so part (a) implies that $\frac{1}{f}$ has a removable singularity at z_0. More precisely, the function $g : D[z_0, R] \to \mathbb{C}$ defined by

$$g(z) := \begin{cases} \frac{1}{f(z)} & \text{if } z \in D[z_0, R], \\ 0 & \text{if } z = z_0, \end{cases}$$

is holomorphic. By Theorem 8.14, there exist a positive integer n and a holomorphic function h on $D[z_0, R]$ such that $h(z_0) \neq 0$ and $g(z) = (z - z_0)^n h(z)$. Actually, $h(z) \neq 0$ for all $z \in D[z_0, R]$ since $g(z) \neq 0$ for all $z \in D[z_0, R]$. Thus

$$\lim_{z \to z_0} (z - z_0)^{n+1} f(z) = \lim_{z \to z_0} \frac{(z - z_0)^{n+1}}{g(z)}$$
$$= \lim_{z \to z_0} \frac{z - z_0}{h(z)} = \frac{1}{h(z_0)} \lim_{z \to z_0} (z - z_0) = 0.$$

Note that $\frac{1}{h}$ is holomorphic and non-zero on $D[z_0, R]$, $n > 0$, and

$$f(z) = \frac{1}{g(z)} = \frac{1}{(z - z_0)^n} \cdot \frac{1}{h(z)} \qquad \text{for all } z \in D[z_0, R].$$

Conversely, suppose z_0 is not removable and $\lim_{z \to z_0} (z - z_0)^{n+1} f(z) = 0$ for some non-negative integer n. We choose the *smallest* such n. By part (a), $h(z) := (z - z_0)^n f(z)$ has a removable singularity at z_0, so there is a holomorphic function

g on $D[z_0, R]$ that agrees with h on $D[z_0, R]$. Now if $n = 0$ this just says that f has a removable singularity at z_0, which we have excluded. Hence $n > 0$. Since n was chosen as small as possible and $n - 1$ is a non-negative integer less than n, we must have $g(z_0) = \lim_{z \to z_0} (z - z_0)^n f(z) \neq 0$. Summarizing, g is holomorphic on $D[z_0, R]$ and non-zero at z_0, $n > 0$, and

$$f(z) = \frac{g(z)}{(z-z_0)^n} \quad \text{for all } z \in D[z_0, R].$$

But then z_0 is a pole of f, since

$$\lim_{z \to z_0} |f(z)| = \lim_{z \to z_0} \left| \frac{h(z)}{(z-z_0)^n} \right| = \lim_{z \to z_0} \left| \frac{g(z)}{(z-z_0)^n} \right|$$
$$= |g(z_0)| \lim_{z \to z_0} \frac{1}{|z-z_0|^n} = \infty. \quad \square$$

We underline one feature of the last part of our proof:

Corollary 9.6. Suppose f is holomorphic in $D[z_0, R]$. Then f has a pole at z_0 if and only if there exist a positive integer m and a holomorphic function $g : D[z_0, R] \to \mathbb{C}$, such that $g(z_0) \neq 0$ and

$$f(z) = \frac{g(z)}{(z-z_0)^m} \quad \text{for all } z \in D[z_0, R].$$

If z_0 is a pole then m is unique.

Proof. The only part not covered in the proof of Theorem 9.5 is uniqueness of m. Suppose $f(z) = (z - z_0)^{-m_1} g_1(z)$ and $f(z) = (z - z_0)^{-m_2} g_2(z)$ both work, with $m_2 > m_1$. Then $g_2(z) = (z - z_0)^{m_2 - m_1} g_1(z)$, and plugging in $z = z_0$ yields $g_2(z_0) = 0$, violating $g_2(z_0) \neq 0$. $\quad \square$

Definition. The integer m in Corollary 9.6 is the *order* of the pole z_0.

This definition, naturally coming out of Corollary 9.6, parallels that of the multiplicity of a zero, which naturally came out of Theorem 8.14. The two results also show that f has a zero at z_0 of multiplicity m if and only if $\frac{1}{f}$ has a pole of order m. We will make use of the notions of zeros and poles quite extensively in this chapter.

CLASSIFICATION OF SINGULARITIES

You might have noticed that the Proposition 9.5 did not include any result on essential singularities. Not only does the next theorem make up for this but it also nicely illustrates the strangeness of essential singularities. To appreciate the following result, we suggest meditating about its statement over a good cup of coffee.

Theorem 9.7 (Casorati[1]–Weierstraß). If z_0 is an essential singularity of f and r is any positive real number, then every $w \in \mathbb{C}$ is arbitrarily close to a point in $f(D[z_0, r])$. That is, for any $w \in \mathbb{C}$ and any $\varepsilon > 0$ there exists $z \in D[z_0, r]$ such that $|w - f(z)| < \varepsilon$.

In the language of topology, Theorem 9.7 says that the image of any punctured disk centered at an essential singularity is *dense* in $\mathbb{C}$.

There is a stronger theorem, beyond the scope of this book, which implies the Casorati–Weierstraß Theorem 9.7. It is due to Charles Emile Picard (1856–1941) and says that the image of any punctured disk centered at an essential singularity misses at most one point of $\mathbb{C}$. (It is worth coming up with examples of functions that do not miss any point in $\mathbb{C}$ and functions that miss exactly one point. Try it!)

Proof. Suppose (by way of contradiction) that there exist $w \in \mathbb{C}$ and $\varepsilon > 0$ such that for all $z \in D[z_0, r]$

$$|w - f(z)| \geq \varepsilon.$$

Then the function $g(z) := \frac{1}{f(z) - w}$ stays bounded as $z \to z_0$, and so

$$\lim_{z \to z_0} \frac{z - z_0}{f(z) - w} = \lim_{z \to z_0} (z - z_0) g(z) = 0.$$

(Proposition 9.5(a) tells us that g has a removable singularity at z_0.) Hence

$$\lim_{z \to z_0} \left| \frac{f(z) - w}{z - z_0} \right| = \infty$$

and so the function $\frac{f(z) - w}{z - z_0}$ has a pole at z_0. By Proposition 9.5(b), there is a positive integer n so that

$$\lim_{z \to z_0} (z - z_0)^{n+1} \frac{f(z) - w}{z - z_0} = \lim_{z \to z_0} (z - z_0)^n (f(z) - w) = 0.$$

---

[1] Felice Casorati (1835–1890).

Invoking Proposition 9.5 again, we conclude that the function $f(z)-w$ has a pole or removable singularity at z_0, which implies the same holds for $f(z)$, a contradiction. □

The following classifies singularities according to their Laurent series, and is very often useful in calculations.

Proposition 9.8. Suppose z_0 is an isolated singularity of f with Laurent series

$$f(z) = \sum_{k \in \mathbb{Z}} c_k (z-z_0)^k,$$

valid in some punctured disk centered at z_0. Then

(a) z_0 is removable if and only if there are no negative exponents (that is, the Laurent series is a power series);

(b) z_0 is a pole if and only if there are finitely many negative exponents, and the order of the pole is the largest k such that $c_{-k} \neq 0$;

(c) z_0 is essential if and only if there are infinitely many negative exponents.

Proof. (a) Suppose z_0 is removable. Then there exists a holomorphic function $g : D[z_0, R] \to \mathbb{C}$ that agrees with f on $\dot{D}[z_0, R]$, for some $R > 0$. By Theorem 8.8, g has a power series expansion centered at z_0, which coincides with the Laurent series of f at z_0, by Corollary 8.25.

Conversely, if the Laurent series of f at z_0 has only nonnegative powers, we can use it to define a function that is holomorphic at z_0.

(b) Suppose z_0 is a pole of order n. Then, by Corollary 9.6, $f(z) = (z-z_0)^{-n} g(z)$ on some punctured disk $\dot{D}[z_0, R]$, where g is holomorphic on $D[z_0, R]$ and $g(z_0) \neq 0$. Thus $g(z) = \sum_{k \geq 0} c_k (z-z_0)^k$ in $D[z_0, R]$ with $c_0 \neq 0$, so

$$f(z) = (z-z_0)^{-n} \sum_{k \geq 0} c_k (z-z_0)^k = \sum_{k \geq -n} c_{k+n} (z-z_0)^k,$$

and this is the Laurent series of f, by Corollary 8.25.

Conversely, suppose that

$$f(z) = \sum_{k \geq -n} c_k (z-z_0)^k = (z-z_0)^{-n} \sum_{k \geq -n} c_k (z-z_0)^{k+n}$$
$$= (z-z_0)^{-n} \sum_{k \geq 0} c_{k-n} (z-z_0)^k,$$

where $c_{-n} \neq 0$. Define $g(z) := \sum_{k \geq 0} c_{k-n}(z-z_0)^k$. Then g is holomorphic at z_0 and $g(z_0) = c_{-n} \neq 0$ so, by Corollary 9.6, f has a pole of order n at z_0.

(c) follows by definition: an essential singularity is neither removable nor a pole. □

Example 9.9. The order of the pole at 0 of $f(z) = \frac{\sin(z)}{z^3}$ is 2 because (by Example 8.4)

$$f(z) = \frac{\sin(z)}{z^3} = \frac{z - \frac{z^3}{3!} + \frac{z^5}{5!} - \cdots}{z^3} = \frac{1}{z^2} - \frac{1}{3!} + \frac{z^2}{5!} - \cdots$$

and the smallest power of z with nonzero coefficient in this series is -2. □

9.2 Residues

We now pick up the thread from Corollary 8.27 and apply it to the Laurent series

$$f(z) = \sum_{k \in \mathbb{Z}} c_k (z-z_0)^k$$

at an isolated singularity z_0 of f. It says that if γ is any positively oriented, simple, closed, piecewise smooth path in the punctured disk of convergence of this Laurent series, and z_0 is inside γ, then

$$\int_\gamma f(z)\, dz = 2\pi i\, c_{-1}.$$

Definition. Suppose z_0 is an isolated singularity of a function f with Laurent series $\sum_{k \in \mathbb{Z}} c_k (z-z_0)^k$. Then c_{-1} is the *residue of f at z_0*, denoted by $\operatorname{Res}_{z=z_0}(f(z))$ or $\operatorname{Res}(f(z), z = z_0)$.

Corollary 8.27 suggests that we can compute integrals over closed curves by finding the residues at isolated singularities, and our next theorem makes this precise.

Theorem 9.10 (Residue Theorem). Suppose f is holomorphic in the region G, except for isolated singularities, and γ is a positively oriented, simple, closed, piecewise smooth path that avoids the singularities of f, and $\gamma \sim_G 0$. Then there are only finitely many singularities inside γ, and

$$\int_\gamma f = 2\pi i \sum_k \operatorname{Res}_{z=z_k}(f(z))$$

where the sum is taken over all singularities z_k inside γ.

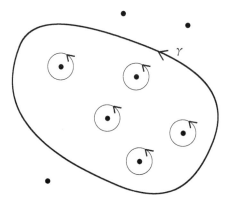

Figure 9.1: Proof of Theorem 9.10.

Proof. First, let S be the set of singularities inside γ. S is closed (since the set of points in G where f is holomorphic is open) and bounded (since the inside of γ is bounded), and the points of S are isolated in S (by Theorem 8.14((b))). An application of Exercise 9.23 shows that S is finite.

Now we follow the approach started in Figure 7.1: as with that integration path, we "subdivide" γ so that we can replace it by closed curves around the singularities inside γ. These curves, in turn, can then be transformed to circles around the singularities, as suggested by Figure 9.1. By Cauchy's Theorem 4.18, $\int_\gamma f$ equals the sum of the integrals of f over these circles. Now use Corollary 8.27. □

RESIDUES

Computing integrals is as easy (or hard!) as computing residues. The following two propositions start the range of tricks you may use when computing residues.

Proposition 9.11. (a) If z_0 is a removable singularity of f then $\operatorname*{Res}_{z=z_0}(f(z)) = 0$.

(b) If z_0 is a pole of f of order n then

$$\operatorname*{Res}_{z=z_0}(f(z)) = \frac{1}{(n-1)!} \lim_{z \to z_0} \frac{d^{n-1}}{dz^{n-1}}\left((z-z_0)^n f(z)\right).$$

Proof. (a) follows from Proposition 9.8(a).

(b) We know by Proposition 9.8(b) that the Laurent series at z_0 looks like

$$f(z) = \sum_{k \geq -n} c_k (z-z_0)^k.$$

But then

$$(z-z_0)^n f(z) = \sum_{k \geq -n} c_k (z-z_0)^{k+n}$$

is a power series, and we can use Taylor's formula (Corollary 8.5) to compute c_{-1}. □

It is worth noting that we are really coming full circle here: compare this proposition to Cauchy's Integral Formulas (Theorems 4.27 & 5.1 and Corollary 8.11).

Example 9.12. The integrand $\frac{\exp(z)}{\sin(z)}$ in Example 8.28 has poles of order 1 at 0 and π. We thus compute

$$\operatorname*{Res}_{z=0}\left(\frac{\exp(z)}{\sin(z)}\right) = \lim_{z \to 0}\left(z \frac{\exp(z)}{\sin(z)}\right) = \exp(0) \lim_{z \to 0} \frac{z}{\sin(z)} = 1$$

and

$$\operatorname*{Res}_{z=\pi}\left(\frac{\exp(z)}{\sin(z)}\right) = \lim_{z \to \pi}\left((z-\pi) \frac{\exp(z)}{\sin(z)}\right) = \exp(\pi) \lim_{z \to \pi} \frac{z-\pi}{\sin(z)} = -e^\pi,$$

confirming our computations in Example 8.28. □'

Example 9.13. Revisiting Example 9.9, the function $f(z) = \frac{\sin(z)}{z^3}$ has a double pole at 0 with

$$\operatorname*{Res}_{z=0}\left(\frac{\sin(z)}{z^3}\right) = \lim_{z \to 0} \frac{d}{dz}\left(z^2 \frac{\sin(z)}{z^3}\right) = \lim_{z \to 0}\left(\frac{z\cos(z) - \sin(z)}{z^2}\right) = 0,$$

after a few iterations of L'Hôpital's Rule. (In this case, it is simpler to read the residue off the Laurent series in Example 9.9.) □

Proposition 9.14. Suppose f and g are holomorphic at z_0, which is a simple zero of g (i.e., a zero of multiplicity 1). Then

$$\operatorname*{Res}_{z=z_0}\left(\frac{f(z)}{g(z)}\right) = \frac{f(z_0)}{g'(z_0)}.$$

Proof. The functions f and g have power series centered at z_0; the one for g has by assumption no constant term:

$$f(z) = \sum_{k\geq 0} a_k(z-z_0)^k$$
$$g(z) = \sum_{k\geq 1} b_k(z-z_0)^k = (z-z_0)\sum_{k\geq 1} b_k(z-z_0)^{k-1}.$$

Let $h(z) := \sum_{k\geq 1} b_k(z-z_0)^{k-1}$ and note that $h(z_0) = b_1 \neq 0$. Hence

$$\frac{f(z)}{g(z)} = \frac{f(z)}{(z-z_0)h(z)},$$

and the function $\frac{f}{h}$ is holomorphic at z_0. By Prop 9.11 and Taylor's formula (Corollary 8.5),

$$\operatorname*{Res}_{z=z_0}\left(\frac{f(z)}{g(z)}\right) = \lim_{z\to z_0}\left((z-z_0)\frac{f(z)}{(z-z_0)h(z)}\right) = \frac{f(z_0)}{h(z_0)} = \frac{a_0}{b_1} = \frac{f(z_0)}{g'(z_0)}. \quad \square$$

Example 9.15. Revisiting once more Example 8.28, we note that $f(z) = \exp(z)$ and $g(z) = \sin(z)$ fit the bill. Thus

$$\operatorname*{Res}_{z=0}\left(\frac{\exp(z)}{\sin(z)}\right) = \frac{\exp(0)}{\cos(0)} = 1$$

and

$$\operatorname*{Res}_{z=\pi}\left(\frac{\exp(z)}{\sin(z)}\right) = \frac{\exp(\pi)}{\cos(\pi)} = -e^{\pi},$$

confirming once more our computations in Examples 8.28 and 9.12. □

Example 9.16. We compute the residue of $\frac{z^2+2}{(\exp(z)-1)\cos(z)}$ at $z_0 = 2\pi i$, by applying Proposition 9.14 with $f(z) = \frac{z^2+2}{\cos(z)}$ and $g(z) = \exp(z) - 1$. Thus

$$\operatorname*{Res}_{z=2\pi i}\left(\frac{z^2+2}{(\exp(z)-1)\cos(z)}\right) = \frac{\frac{(2\pi i)^2+2}{\cos(2\pi i)}}{\exp(2\pi i)} = \frac{-4\pi^2+2}{\cosh(2\pi)}. \qquad \square$$

An extension of Proposition 9.14 of sorts is given in Exercise 9.12.

9.3 Argument Principle and Rouché's Theorem

In the previous section we saw how to compute integrals via residues, but in many applications we actually do not have an explicit expression for a function that we need to integrate (or this expression is very complicated). However, it may still be possible to compute the value of a function at any given point. In this situation we cannot immediately apply the Residue Theorem because we don't know where the singularities are. Of course, we could use numerical integration to compute integrals over any path, but computationally this task could be very resource intensive. But if we do know the singularities, we can compute the residues numerically by computing a finite number of the integrals over small circles around these singularities. And after that we can apply the residue theorem to compute the integral over any closed path very effectively: we just sum up the residues inside this path. The argument principle that we study below, in particular, addresses this question. We start by introducing the logarithmic derivative.

Suppose we have a differentiable function f. Differentiating $\operatorname{Log} f$ (where Log is some branch of the logarithm) gives $\frac{f'}{f}$, which is one good reason to call this quotient the *logarithmic derivative* of f. It has some remarkable properties, one of which we would like to discuss here.

Now let's say we have two functions f and g holomorphic in some region. Then the logarithmic derivative of their product behaves very nicely:

$$\frac{(fg)'}{fg} = \frac{f'g+fg'}{fg} = \frac{f'}{f} + \frac{g'}{g}.$$

We can apply this fact to the following situation: Suppose that f is holomorphic in a region G and f has (finitely many) zeros $z_1, \ldots, z_j$ of multiplicities $n_1, \ldots, n_j$, respectively. By Theorem 8.14, we can express f as

$$f(z) = (z - z_1)^{n_1} \cdots (z - z_j)^{n_j} g(z),$$

where g is also holomorphic in G and never zero. Let's compute the logarithmic derivative of f and play the same remarkable cancellation game as above:

$$\frac{f'(z)}{f(z)} =$$
$$\frac{n_1(z-z_1)^{n_1-1}(z-z_2)^{n_2}\cdots(z-z_j)^{n_j}g(z) + \cdots + (z-z_1)^{n_1}\cdots(z-z_j)^{n_j}g'(z)}{(z-z_1)^{n_1}\cdots(z-z_j)^{n_j}g(z)}$$

$$= \frac{n_1}{z-z_1} + \frac{n_2}{z-z_2} + \cdots + \frac{n_j}{z-z_j} + \frac{g'(z)}{g(z)}. \quad (9.1)$$

Something similar happens if f has finitely many poles in G. In Exercise 9.19, we invite you to prove that, if $p_1, \ldots, p_k$ are all the poles of f in G with order $m_1, \ldots, m_k$, respectively, then the logarithmic derivative of f can be expressed as

$$\frac{f'(z)}{f(z)} = -\frac{m_1}{z-p_1} - \frac{m_2}{z-p_2} - \cdots - \frac{m_k}{z-p_k} + \frac{g'(z)}{g(z)}, \quad (9.2)$$

where g is a function without poles in G. Naturally, we can combine the expressions for zeros and poles, as we will do in a moment.

Definition. A function f is *meromorphic* in the region G if f is holomorphic in G except for poles.

Theorem 9.17 (Argument Principle[2]). Suppose f is meromorphic in a region G and γ is a positively oriented, simple, closed, piecewise smooth path that does not pass through any zero or pole of f, and $\gamma \sim_G 0$. Denote by $Z(f, \gamma)$ the number of zeros of f inside γ counted according to multiplicity and by $P(f, \gamma)$ the number of poles of f inside γ counted according to order. Then

$$\frac{1}{2\pi i} \int_\gamma \frac{f'}{f} = Z(f, \gamma) - P(f, \gamma).$$

---

[2] The name *Argument Principle* stems from interpreting the integral $\int_\gamma \frac{f'}{f}$ as the change in the argument of $f(z)$ as z traverses γ, since $\text{Log}(f(z))' = \frac{f'(z)}{f(z)}$.

Proof. Suppose the zeros of f inside γ are $z_1, \ldots, z_j$ of multiplicities $n_1, \ldots, n_j$, respectively, and the poles inside γ are $p_1, \ldots, p_k$ with order $m_1, \ldots, m_k$, respectively. (You may meditate about the fact why there can be only finitely many zeros and poles inside γ.) In fact, we may shrink G, if necessary, so that these are the only zeros and poles in G. By (9.1) and (9.2),

$$\frac{f'(z)}{f(z)} = \frac{n_1}{z - z_1} + \cdots + \frac{n_j}{z - z_j} - \frac{m_1}{z - p_1} - \cdots - \frac{m_k}{z - p_k} + \frac{g'(z)}{g(z)},$$

where g is a function that is holomorphic in G (in particular, without poles) and never zero. Thanks to Cauchy's Theorem 4.18 and Exercise 4.4, the integral is easy:

$$\int_\gamma \frac{f'}{f} = n_1 \int_\gamma \frac{dz}{z - z_1} + \cdots + n_j \int_\gamma \frac{dz}{z - z_j}$$
$$- m_1 \int_\gamma \frac{dz}{z - p_1} - \cdots - m_k \int_\gamma \frac{dz}{z - p_k} + \int_\gamma \frac{g'}{g}$$
$$= 2\pi i \left(n_1 + \cdots + n_j - m_1 - \cdots - m_k \right) + \int_\gamma \frac{g'}{g}.$$

Finally, $\frac{g'}{g}$ is holomorphic in G (because g is never zero in G), so that Corollary 4.20 gives

$$\int_\gamma \frac{g'}{g} = 0. \qquad \square$$

As mentioned above, this beautiful theorem helps to locate poles and zeroes of a function f. The idea is simple: we can first numerically integrate $\frac{f'}{f}$ over a big circle γ that includes all possible paths over which we potentially will be integrating f. Then the numerical value of $\frac{1}{2\pi i} \int_\gamma \frac{f'}{f}$ will be close to an integer that, according to the Argument Principle, equals $Z(f, \gamma) - P(f, \gamma)$. Then we can integrate $\frac{f'}{f}$ over a smaller closed path γ_1 that encompasses half of the interior of γ and find $Z(f, \gamma_1) - P(f, \gamma_1)$. Continuing this process for smaller and smaller regions will (after certain verification) produce small regions where f has exactly one zero or exactly one pole. Integrating f over the boundaries of those small regions that contain poles and dividing by $2\pi i$ gives all residues of f.

Another nice related application of the Argument Principle is a famous theorem due to Eugene Rouché (1832–1910).

Theorem 9.18 (Rouché's Theorem). Suppose f and g are holomorphic in a region G and γ is a positively oriented, simple, closed, piecewise smooth path, such that $\gamma \sim_G 0$ and $|f(z)| > |g(z)|$ for all $z \in \gamma$. Then

$$Z(f+g, \gamma) = Z(f, \gamma).$$

This theorem is of surprising practicality. It allows us to locate the zeros of a function fairly precisely. Here is an illustration.

Example 9.19. All the roots of the polynomial $p(z) = z^5 + z^4 + z^3 + z^2 + z + 1$ have modulus less than two.[3] To see this, let $f(z) = z^5$ and $g(z) = z^4 + z^3 + z^2 + z + 1$. Then for $z \in C[0, 2]$

$$|g(z)| \leq |z|^4 + |z|^3 + |z|^2 + |z| + 1 = 16 + 8 + 4 + 2 + 1 = 31 < 32 = |z|^5 = |f(z)|.$$

So g and f satisfy the condition of the Theorem 9.18. But f has just one root, of multiplicity 5 at the origin, whence

$$Z(p, C[0,2]) = Z(f+g, C[0,2]) = Z(f, C[0,2]) = 5. \qquad \square$$

Proof of Theorem 9.18. By (9.1) and the Argument Principle (Theorem 9.17)

$$Z(f+g, \gamma) = \frac{1}{2\pi i} \int_\gamma \frac{(f+g)'}{f+g} = \frac{1}{2\pi i} \int_\gamma \frac{\left(f\left(1+\frac{g}{f}\right)\right)'}{f\left(1+\frac{g}{f}\right)}$$

$$= \frac{1}{2\pi i} \int_\gamma \left(\frac{f'}{f} + \frac{\left(1+\frac{g}{f}\right)'}{1+\frac{g}{f}} \right) = Z(f, \gamma) + \frac{1}{2\pi i} \int_\gamma \frac{\left(1+\frac{g}{f}\right)'}{1+\frac{g}{f}}.$$

We are assuming that $\left|\frac{g}{f}\right| < 1$ on γ, which means that the function $1 + \frac{g}{f}$ evaluated on γ stays away from $\mathbb{R}_{\leq 0}$. But then $\mathrm{Log}(1 + \frac{g}{f})$ is a well-defined holomorphic function on γ. Its derivative is

$$\frac{\left(1+\frac{g}{f}\right)'}{1+\frac{g}{f}}$$

---

[3] The Fundamental Theorem of Algebra (Theorem 5.11) asserts that p has five roots in $\mathbb{C}$. What's special about the statement of Example 9.19 is that they all have modulus < 2. Note also that there is no general formula for computing roots of a polynomial of degree 5. (Although for this p it's not hard to find one root—and therefore all of them.)

which implies by Corollary 4.13 that

$$\frac{1}{2\pi i}\int_\gamma \frac{\left(1+\frac{g}{f}\right)'}{1+\frac{g}{f}} = 0. \qquad \square$$

Exercises

9.1. Suppose that f has a zero of multiplicity m at a. Explain why $\frac{1}{f}$ has a pole of order m at a.

9.2. Find the poles of the following functions and determine their orders:

(a) $(z^2+1)^{-3}(z-1)^{-4}$ (c) $z^{-5}\sin(z)$ (e) $\dfrac{z}{1-\exp(z)}$

(b) $z\cot(z)$ (d) $\dfrac{1}{1-\exp(z)}$

9.3. Show that if f has an essential singularity at z_0 then $\frac{1}{f}$ also has an essential singularity at z_0.

9.4. Suppose f is a nonconstant entire function. Prove that any complex number is arbitrarily close to a number in $f(\mathbb{C})$. (*Hint*: If f is not a polynomial, use Theorem 9.7 for $f(\frac{1}{z})$.)

9.5. Evaluate the following integrals for $\gamma = C[0,3]$.

(a) $\displaystyle\int_\gamma \cot(z)\,dz$ (d) $\displaystyle\int_\gamma z^2 \exp(\tfrac{1}{z})\,dz$

(b) $\displaystyle\int_\gamma z^3 \cos(\tfrac{3}{z})\,dz$ (e) $\displaystyle\int_\gamma \frac{\exp(z)}{\sinh(z)}\,dz$

(c) $\displaystyle\int_\gamma \frac{dz}{(z+4)(z^2+1)}$ (f) $\displaystyle\int_\gamma \frac{i^{z+4}}{(z^2+16)^2}\,dz$

9.6. Suppose f has a simple pole (i.e., a pole of order 1) at z_0 and g is holomorphic at z_0. Prove that

$$\operatorname*{Res}_{z=z_0}(f(z)g(z)) = g(z_0)\operatorname*{Res}_{z=z_0}(f(z)).$$

9.7. Find the residue of each function at 0:

(a) $z^{-3}\cos(z)$

(b) $\csc(z)$

(c) $\dfrac{z^2+4z+5}{z^2+z}$

(d) $\exp(1-\frac{1}{z})$

(e) $\dfrac{\exp(4z)-1}{\sin^2(z)}$

9.8. Use residues to evaluate the following integrals:

(a) $\displaystyle\int_{C[i-1,1]} \dfrac{dz}{z^4+4}$

(b) $\displaystyle\int_{C[i,2]} \dfrac{dz}{z(z^2+z-2)}$

(c) $\displaystyle\int_{C[0,2]} \dfrac{\exp(z)}{z^3+z}\,dz$

(d) $\displaystyle\int_{C[0,1]} \dfrac{dz}{z^2\sin z}$

(e) $\displaystyle\int_{C[0,3]} \dfrac{\exp(z)}{(z+2)^2\sin z}\,dz$

(f) $\displaystyle\int_{C[\pi,1]} \dfrac{\exp(z)}{\sin(z)\cos(z)}\,dz$

9.9. Use the Residue Theorem 9.10 to re-prove Cauchy's Integral Formulas (Theorems 4.27 & 5.1 and Corollary 8.11).

9.10. Revisiting Exercise 8.34, show that if f is even then $\operatorname{Res}_{z=0}(f(z))=0$.

9.11. Suppose f has an isolated singularity at z_0.

(a) Show that f' also has an isolated singularity at z_0.

(b) Find $\operatorname{Res}_{z=z_0}(f')$.

9.12. Extend Proposition 9.14 by proving, if f and g are holomorphic at z_0, which is a double zero of g, then

$$\operatorname*{Res}_{z=z_0}\left(\dfrac{f(z)}{g(z)}\right) = \dfrac{6f'(z_0)\,g''(z_0)-2f(z_0)\,g'''(z_0)}{3\,g''(z_0)^2}.$$

9.13. Compute $\displaystyle\int_{C[2,3]} \dfrac{\cos(z)}{\sin^2(z)}\,dz$.

9.14. Generalize Example 5.14 and Exercise 5.18 as follows: Let $p(x)$ and $q(x)$ be polynomials such that $q(x) \neq 0$ for $x \in \mathbb{R}$ and the degree of $q(x)$ is at least two larger than the degree of $p(x)$. Prove that $\int_{-\infty}^{\infty} \frac{p(x)}{q(x)} dx$ equals $2\pi i$ times the sum of the residues of $\frac{p(z)}{q(z)}$ at all poles in the upper half plane.

9.15. Compute $\int_{-\infty}^{\infty} \frac{dx}{(1+x^2)^2}$.

9.16. Generalize Exercise 5.19 by deriving conditions under which we can compute $\int_{-\infty}^{\infty} \frac{p(x)\cos(x)}{q(x)} dx$ for polynomials $p(x)$ and $q(x)$, and give a formula for this integral along the lines of Exercise 9.14.

9.17. Compute $\int_{-\infty}^{\infty} \frac{\cos(x)}{1+x^4} dx$.

9.18. Suppose f is entire and $a, b \in \mathbb{C}$ with $a \neq b$ and $|a|, |b| < R$. Evaluate

$$\int_{C[0,R]} \frac{f(z)}{(z-a)(z-b)} dz$$

and use this to give an alternate proof of Liouville's Theorem 5.13. (*Hint*: Show that if f is bounded then the above integral goes to zero as R increases.)

9.19. Prove (9.2).

9.20. Suppose f is meromorphic in the region G, g is holomorphic in G, and γ is a positively oriented, simple, closed, piecewise smooth path that does not pass through any zero or pole of f, and $\gamma \sim_G 0$. Denote the zeros and poles of f inside γ by $z_1, \ldots, z_j$ and $p_1, \ldots, p_k$, respectively, counted according to multiplicity. Prove that

$$\frac{1}{2\pi i} \int_\gamma g \frac{f'}{f} = \sum_{m=1}^{j} g(z_m) - \sum_{n=1}^{k} g(p_n).$$

9.21. Find the number of zeros of

(a) $3\exp(z) - z$ in $\overline{D}[0,1]$

(b) $\frac{1}{3}\exp(z) - z$ in $\overline{D}[0,1]$

(c) $z^4 - 5z + 1$ in $\{z \in \mathbb{C} : 1 \leq |z| \leq 2\}$

9.22. Give another proof of the Fundamental Theorem of Algebra (Theorem 5.11), using Rouché's Theorem 9.18. (*Hint*: If $p(z) = a_n z^n + a_{n-1} z^{n-1} + \cdots + a_1 z + 1$, let $f(z) = a_n z^n$ and $g(z) = a_{n-1} z^{n-1} + a_{n-2} z^{n-2} + \cdots + a_1 z + 1$, and choose as γ a circle that is large enough to make the condition of Rouché's theorem work. You might want to first apply Proposition 5.10 to $g(z)$.)

9.23. Suppose $S \subset \mathbb{C}$ is closed and bounded and all points of S are isolated points of S. Show that S is finite, as follows:

(a) For each $z \in S$ we can choose $\varphi(z) > 0$ so that $D[z, \varphi(z)]$ contains no points of S except z. Show that φ is continuous. (*Hint*: This is really easy if you use the *first* definition of continuity in Section 2.1.)

(b) Assume S is non-empty. By the Extreme Value Theorem A.1, φ has a minimum value, $r_0 > 0$. Let $r = r_0/2$. Since S is bounded, it lies in a disk $D[0, M]$ for some M. Show that the small disks $D[z, r]$, for $z \in S$, are disjoint and lie in $D[0, M + r]$.

(c) Find a bound on the number of such small disks. (*Hint*: Compare the areas of $D[z, r]$ and $D[0, M + r]$.)

Chapter 10

Discrete Applications of the Residue Theorem

All means (even continuous) sanctify the discrete end.
Doron Zeilberger

On the surface, this chapter is just a collection of exercises. They are more involved than any of the ones we've given so far at the end of each chapter, which is one reason why we will lead you through each of the following ones step by step. On the other hand, these sections should really be thought of as a continuation of the book, just in a different format. All of the following problems are of a *discrete* mathematical nature, and we invite you to solve them using *continuous* methods—namely, complex integration. There are very few results in mathematics that so intimately combine discrete and continuous mathematics as does the Residue Theorem 9.10.

10.1 Infinite Sums

In this exercise, we evaluate the sums $\sum_{k \geq 1} \frac{1}{k^2}$ and $\sum_{k \geq 1} \frac{(-1)^k}{k^2}$. We hope the idea how to compute such sums in general will become clear.

(1) Consider the function $f(z) = \frac{\pi \cot(\pi z)}{z^2}$. Compute the residues at all the singularities of f.

(2) Let N be a positive integer and γ_N be the rectangular path from $N + \frac{1}{2} - iN$ to $N + \frac{1}{2} + iN$ to $-N - \frac{1}{2} + iN$ to $-N - \frac{1}{2} - iN$ back to $N + \frac{1}{2} - iN$.

  (a) Show that $|\cot(\pi z)| < 2$ for $z \in \gamma_N$. (*Hint*: Use Exercise 3.36.)

  (b) Show that $\lim_{N \to \infty} \int_{\gamma_N} f = 0$.

(3) Use the Residue Theorem 9.10 to arrive at an identity for $\sum_{k \in \mathbb{Z} \setminus \{0\}} \frac{1}{k^2}$.

189

(4) Evaluate $\sum_{k\geq 1} \frac{1}{k^2}$.

(5) Repeat the exercise with the function $f(z) = \frac{\pi}{z^2 \sin(\pi z)}$ to arrive at an evaluation of
$$\sum_{k\geq 1} \frac{(-1)^k}{k^2}.$$

(*Hint*: To bound this function, you may use the fact that $\frac{1}{\sin^2(z)} = 1 + \cot^2(z)$.)

(6) Evaluate $\sum_{k\geq 1} \frac{1}{k^4}$ and $\sum_{k\geq 1} \frac{(-1)^k}{k^4}$.

We remark that, in the language of Example 7.21, you have computed the evaluations $\zeta(2)$ and $\zeta(4)$ of the Riemann zeta function. The function $\zeta^*(z) := \sum_{k\geq 1} \frac{(-1)^k}{k^z}$ is called the *alternating zeta function*.

10.2 Binomial Coefficients

The binomial coefficient $\binom{n}{k}$ is a natural candidate for being explored analytically, as the *binomial theorem*
$$(x+y)^n = \sum_{k=0}^{n} \binom{n}{k} x^k y^{n-k}$$
(for $x, y \in \mathbb{C}$ and $n \in \mathbb{Z}_{\geq 0}$) tells us that $\binom{n}{k}$ is the coefficient of z^k in $(z+1)^n$. You will derive two sample identities in the course of the exercises below.

(1) Convince yourself that
$$\binom{n}{k} = \frac{1}{2\pi i} \int_\gamma \frac{(z+1)^n}{z^{k+1}} dz$$
where γ is any simple closed piecewise smooth path such that 0 is inside γ.

(2) Derive a recurrence relation for binomial coefficients from the fact that $\frac{1}{z} + 1 = \frac{z+1}{z}$. (*Hint*: Multiply both sides by $\frac{(z+1)^n}{z^k}$.)

(3) Now suppose $x \in \mathbb{R}$ with $|x| < 1/4$. Find a simple closed path γ surrounding the origin such that
$$\sum_{k\geq 0} \left(\frac{(z+1)^2}{z} x \right)^k$$
converges uniformly on γ as a function of z. Evaluate this sum.

(4) Keeping x and γ from ((3)), convince yourself that

$$\sum_{k\geq 0} \binom{2k}{k} x^k = \frac{1}{2\pi i} \sum_{k\geq 0} \int_\gamma \frac{(z+1)^{2k}}{z^{k+1}} x^k\, dz,$$

use ((3)) to interchange summation and integral, and use the Residue Theorem 9.10 to evaluate the integral, giving an identity for $\sum_{k\geq 0} \binom{2k}{k} x^k$.

10.3 Fibonacci Numbers

The *Fibonacci*[1] *numbers* are a sequence of integers defined recursively through

$$f_0 = 0$$
$$f_1 = 1$$
$$f_n = f_{n-1} + f_{n-2} \qquad \text{for } n \geq 2.$$

Let $F(z) = \sum_{k\geq 0} f_n z^n$.

(1) Show that F has a positive radius of convergence.

(2) Show that the recurrence relation among the f_n implies that $F(z) = \frac{z}{1-z-z^2}$. (*Hint*: Write down the power series of $z F(z)$ and $z^2 F(z)$ and rearrange both so that you can easily add.)

(3) Verify that

$$\operatorname*{Res}_{z=0}\left(\frac{1}{z^n(1-z-z^2)}\right) = f_n.$$

(4) Use the Residue Theorem 9.10 to derive an identity for f_n. (*Hint*: Integrate

$$\frac{1}{z^n(1-z-z^2)}$$

around $C[0, R]$ and show that this integral vanishes as $R \to \infty$.)

---
[1] Named after Leonardo Pisano Fibonacci (1170–1250).

(5) Generalize to other sequences defined by recurrence relations, e.g., the *Tribonacci numbers*

$$t_0 = 0$$
$$t_1 = 0$$
$$t_2 = 1$$
$$t_n = t_{n-1} + t_{n-2} + t_{n-3} \qquad \text{for } n \geq 3.$$

10.4 The Coin-Exchange Problem

In this exercise, we will solve and extend a classical problem of Ferdinand Georg Frobenius (1849–1917). Suppose a and b are relatively prime[2] positive integers, and suppose t is a positive integer. Consider the function

$$f(z) = \frac{1}{(1-z^a)(1-z^b) z^{t+1}}.$$

(1) Compute the residues at all nonzero poles of f.

(2) Verify that $\operatorname{Res}_{z=0}(f) = N(t)$, where

$$N(t) = |\{(m,n) \in \mathbb{Z} : m, n \geq 0, \ ma + nb = t\}|.$$

(3) Use the Residue Theorem, Theorem 9.10, to derive an identity for $N(t)$. (*Hint*: Integrate f around $C[0, R]$ and show that this integral vanishes as $R \to \infty$.)

(4) Use the following three steps to simplify this identity to

$$N(t) = \frac{t}{ab} - \left\{ \frac{b^{-1}t}{a} \right\} - \left\{ \frac{a^{-1}t}{b} \right\} + 1.$$

---

[2] This means that the integers do not have any common factor.

Here, $\{x\}$ denotes the fractional part[3] of x, $a^{-1}a \equiv 1 \pmod{b}$[4], and $b^{-1}b \equiv 1 \pmod{a}$.

(a) Verify that for $b = 1$,

$$N(t) = |\{(m,n) \in \mathbb{Z}: m, n \geq 0, ma + n = t\}| = |\{m \in \mathbb{Z}: m \geq 0, ma \leq t\}|$$
$$= \left|\left[0, \frac{t}{a}\right] \cap \mathbb{Z}\right| = \frac{t}{a} - \left\{\frac{t}{a}\right\} + 1.$$

(b) Use this together with the identity found in ((3)) to obtain

$$\frac{1}{a} \sum_{k=1}^{a-1} \frac{1}{(1 - e^{2\pi i k/a}) e^{2\pi i k t/a}} = -\left\{\frac{t}{a}\right\} + \frac{1}{2} - \frac{1}{2a}.$$

(c) Verify that

$$\sum_{k=1}^{a-1} \frac{1}{(1 - e^{2\pi i k b/a}) e^{2\pi i k t/a}} = \sum_{k=1}^{a-1} \frac{1}{(1 - e^{2\pi i k/a}) e^{2\pi i k b^{-1} t/a}}.$$

(5) Prove that $N(ab - a - b) = 0$, and $N(t) > 0$ for all $t > ab - a - b$.

Historical remark. Given relatively prime positive integers $a_1, a_2, \ldots, a_n$, let's call an integer t *representable* if there exist nonnegative integers $m_1, m_2, \ldots, m_n$ such that

$$t = m_1 a_1 + m_2 a_2 + \cdots + m_n a_n.$$

(There are many scenarios in which you may ask whether or not t is representable, given fixed $a_1, a_2, \ldots, a_n$; for example, if the a_j's are coin denomination, this question asks whether you can give exact change for t.) In the late 19th century, Frobenius raised the problem of finding the largest integer that is not representable. We call this largest integer the *Frobenius number* $g(a_1, \ldots, a_n)$. It is well known (probably at least since the 1880's, when James Joseph Sylvester (1814–1897) studied the Frobenius problem) that $g(a_1, a_2) = a_1 a_2 - a_1 - a_2$. You verified this result in ((5)). For $n > 2$, there is no nice closed formula for $g(a_1, \ldots, a_n)$. The formula in ((4))

---
[3] The *fractional part* of a real number x is, loosely speaking, the part after the decimal point. More thoroughly, the *greatest integer function* of x, denoted by $\lfloor x \rfloor$, is the greatest integer not exceeding x. The fractional part is then $\{x\} = x - \lfloor x \rfloor$.
[4] This means that a^{-1} is an integer such that $a^{-1}a = 1 + kb$ for some $k \in \mathbb{Z}$.

is due to Tiberiu Popoviciu (1906–1975), though an equivalent version of it was already known to Peter Barlow (1776–1862).

10.5 Dedekind Sums

This exercise outlines one more nontraditional application of the Residue Theorem 9.10. Given two positive, relatively prime integers a and b, let

$$f(z) := \cot(\pi a z) \cot(\pi b z) \cot(\pi z).$$

(1) Choose an $\varepsilon > 0$ such that the rectangular path γ_R from $1-\varepsilon-iR$ to $1-\varepsilon+iR$ to $-\varepsilon+iR$ to $-\varepsilon-iR$ back to $1-\varepsilon-iR$ does not pass through any of the poles of f.

   (a) Compute the residues for the poles of f inside γ_R. *Hint:* Use the periodicity of the cotangent and the fact that

   $$\cot z = \frac{1}{z} - \frac{1}{3} z + \text{higher-order terms}.$$

   (b) Prove that $\lim_{R \to \infty} \int_{\gamma_R} f = -2i$ and deduce that for any $R > 0$

   $$\int_{\gamma_R} f = -2i.$$

(2) Define

$$s(a,b) := \frac{1}{4b} \sum_{k=1}^{b-1} \cot\left(\frac{\pi k a}{b}\right) \cot\left(\frac{\pi k}{b}\right). \tag{10.1}$$

Use the Residue Theorem 9.10 to show that

$$s(a,b) + s(b,a) = -\frac{1}{4} + \frac{1}{12}\left(\frac{a}{b} + \frac{1}{ab} + \frac{b}{a}\right). \tag{10.2}$$

(3) Generalize (10.1) and (10.2).

Historical remark. The sum in (10.1) is called a *Dedekind[5] sum*. It first appeared in the study of the *Dedekind η-function*

$$\eta(z) = \exp\left(\tfrac{\pi i z}{12}\right) \prod_{k \geq 1} (1 - \exp(2\pi i k z))$$

in the 1870's and has since intrigued mathematicians from such different areas as topology, number theory, and discrete geometry. The *reciprocity law* (10.2) is the most important and famous identity of the Dedekind sum. The proof that is outlined here is due to Hans Rademacher (1892–1969).

---

[5]Named after Julius Wilhelm Richard Dedekind (1831–1916).

Appendix: Theorems from Calculus

Here we collect a few theorems from real calculus that we make use of in the course of the text.

Theorem A.1 (Extreme-Value Theorem). Suppose $K \subset \mathbb{R}^n$ is closed and bounded and $f: K \to \mathbb{R}$ is continuous. Then f has a minimum and maximum value, i.e.,

$$\min_{x \in K} f(x) \quad \text{and} \quad \max_{x \in K} f(x)$$

exist in $\mathbb{R}$.

Theorem A.2 (Mean-Value Theorem). Suppose $I \subseteq \mathbb{R}$ is an interval, $f: I \to \mathbb{R}$ is differentiable, and $x, x + \Delta x \in I$. Then there exists $0 < a < 1$ such that

$$\frac{f(x + \Delta x) - f(x)}{\Delta x} = f'(x + a\Delta x).$$

Many of the most important results of analysis concern combinations of limit operations. The most important of all calculus theorems combines differentiation and integration (in two ways):

Theorem A.3 (Fundamental Theorem of Calculus). Suppose $f: [a, b] \to \mathbb{R}$ is continuous.

(a) The function $F: [a, b] \to \mathbb{R}$ defined by $F(x) = \int_a^x f(t)\, dt$ is differentiable and $F'(x) = f(x)$.

(b) If F is any antiderivative of f, that is, $F' = f$, then $\int_a^b f(x)\, dx = F(b) - F(a)$.

Theorem A.4. If $f, g: [a, b] \to \mathbb{R}$ are continuous functions and $c \in \mathbb{R}$ then

$$\int_a^b \bigl(f(x) + c\, g(x)\bigr)\, dx = \int_a^b f(x)\, dx + c \int_a^b g(x)\, dx.$$

Theorem A.5. If $f, g: [a, b] \to \mathbb{R}$ are continuous functions then

$$\left| \int_a^b f(x) g(x)\, dx \right| \leq \int_a^b |f(x) g(x)|\, dx \leq \left(\max_{a \leq x \leq b} |f(x)| \right) \int_a^b |g(x)|\, dx.$$

Theorem A.6. If $g : [a, b] \to \mathbb{R}$ is differentiable, g' is continuous, and $f : [g(a), g(b)] \to \mathbb{R}$ is continuous then

$$\int_a^b f(g(t)) g'(t) \, dt = \int_{g(a)}^{g(b)} f(x) \, dx.$$

For functions of several variables we can perform differentiation/integration operations in any order, if we have sufficient continuity:

Theorem A.7. If the mixed partials $\frac{\partial^2 f}{\partial x \partial y}$ and $\frac{\partial^2 f}{\partial y \partial x}$ are defined on an open set $G \subseteq \mathbb{R}^2$ and are continuous at a point $(x_0, y_0) \in G$, then they are equal at (x_0, y_0).

Theorem A.8. If f is continuous on $[a, b] \times [c, d] \subset \mathbb{R}^2$ then

$$\int_a^b \int_c^d f(x, y) \, dy \, dx = \int_c^d \int_a^b f(x, y) \, dx \, dy.$$

We can apply differentiation and integration with respect to different variables in either order:

Theorem A.9 (Leibniz's Rule[1]). Suppose f is continuous on $[a, b] \times [c, d] \subset \mathbb{R}^2$ and the partial derivative $\frac{\partial f}{\partial x}$ exists and is continuous on $[a, b] \times [c, d]$. Then

$$\frac{d}{dx} \int_c^d f(x, y) \, dy = \int_c^d \frac{\partial f}{\partial x}(x, y) \, dy.$$

Leibniz's Rule follows from the Fundamental Theorem of Calculus (Theorem A.3). You can try to prove it, e.g., as follows: Define $F(x) = \int_c^d f(x, y) \, dy$, get an expression for $F(x) - F(a)$ as an iterated integral by writing $f(x, y) - f(a, y)$ as the integral of $\frac{\partial f}{\partial x}$, interchange the order of integrations, and then differentiate using Theorem A.3.

Theorem A.10 (Green's Theorem[2]). Let C be a positively oriented, piecewise smooth, simple, closed path in $\mathbb{R}^2$ and let D be the set bounded by C. If $f(x, y)$ and $g(x, y)$ have continuous partial derivatives on an open region containing D then

$$\int_C f \, dx + g \, dy = \int_D \frac{\partial f}{\partial x} - \frac{\partial g}{\partial y} \, dx \, dy.$$

---
[1] Named after Gottfried Wilhelm Leibniz (1646–1716).
[2] Named after George Green (1793–1841).

Theorem A.11 (L'Hôspital's Rule[3]). Suppose $I \subset \mathbb{R}$ is an open interval and either c is in I or c is an endpoint of I. Suppose f and g are differentiable functions on $I \setminus \{c\}$ with $g'(x)$ never zero. Suppose

$$\lim_{x \to c} f(x) = 0, \quad \lim_{x \to c} g(x) = 0, \quad \lim_{x \to c} \frac{f'(x)}{g'(x)} = L.$$

Then

$$\lim_{x \to c} \frac{f(x)}{g(x)} = L.$$

There are many extensions of L'Hôspital's rule. In particular, the rule remains true if any of the following changes are made:

- L is infinite.

- I is unbounded and c is an infinite endpoint of I.

- $\lim_{x \to c} f(x)$ and $\lim_{x \to c} g(x)$ are both infinite.

---

[3] Named after Guillaume de l'Hôspital (1661–1704).

Solutions to Selected Exercises

1.1 (a) $7-i$
   (b) $1-i$
   (c) $-11-2i$
   (d) 5
   (e) $-2+3i$

1.2 (b) $\frac{19}{25} - \frac{8}{25}i$
   (c) 1

1.3 (a) $\sqrt{5}$, $-2-i$
   (b) $5\sqrt{5}$, $5-10i$
   (c) $\sqrt{\frac{10}{11}}$, $\frac{3}{11}(\sqrt{2}-1) + \frac{i}{11}(\sqrt{2}+9)$
   (d) 8, $8i$

1.4 (a) $2\,e^{i\frac{\pi}{2}}$
   (b) $\sqrt{2}\,e^{i\frac{\pi}{4}}$
   (c) $2\sqrt{3}i\,e^{i\frac{5\pi}{6}}$
   (d) $e^{i\frac{3\pi}{2}}$

1.5 (a) $-1+i$
   (b) $34i$
   (c) -1
   (d) 2

1.9 $\pm e^{i\frac{\pi}{4}} - 1$

1.11 (a) $z = e^{i\frac{\pi}{3}k}$, $k = 0, 1, \ldots, 5$
   (b) $z = 2\,e^{i\frac{\pi}{4} + i\frac{\pi}{2}k}$, $k = 0, 1, 2, 3$

1.18 $\cos\frac{\pi}{5} = \frac{1}{4}(\sqrt{5}+1)$ and $\cos\frac{2\pi}{5} = \frac{1}{4}(\sqrt{5}-1)$.

2.2 (a) 0
   (b) $1+i$

2.17 (a) differentiable and holomorphic in $\mathbb{C}$ with derivative $-e^{-x}e^{-iy}$
 (b) nowhere differentiable or holomorphic
 (c) differentiable only on $\{x+iy \in \mathbb{C}: x=y\}$ with derivative $2x$, nowhere holomorphic
 (d) nowhere differentiable or holomorphic
 (e) differentiable and holomorphic in $\mathbb{C}$ with derivative
 $-\sin x \cosh y - i \cos x \sinh y$
 (f) nowhere differentiable or holomorphic
 (g) differentiable only at 0 with derivative 0, nowhere holomorphic
 (h) differentiable only at 0 with derivative 0, nowhere holomorphic
 (i) differentiable only at i with derivative i, nowhere holomorphic
 (j) differentiable and holomorphic in $\mathbb{C}$ with derivative $2y - 2xi = -2iz$
 (k) differentiable only at 0 with derivative 0, nowhere holomorphic
 (l) differentiable only at 0 with derivative 0, nowhere holomorphic

2.24 (a) $2xy$
 (b) $\cos(x)\sinh(y)$

3.44 (a) differentiable at 0, nowhere holomorphic
 (b) differentiable and holomorphic on $\mathbb{C} \setminus \{-1, e^{i\frac{\pi}{3}}, e^{-i\frac{\pi}{3}}\}$
 (c) differentiable and holomorphic on $\mathbb{C} \setminus \{x+iy \in \mathbb{C}: x \geq -1, y = 2\}$
 (d) nowhere differentiable or holomorphic
 (e) differentiable and holomorphic on $\mathbb{C} \setminus \{x+iy \in \mathbb{C}: x \leq 3, y = 0\}$
 (f) differentiable and holomorphic in $\mathbb{C}$ (i.e. entire)

3.45 (a) $z = i$
 (b) there is no solution
 (c) $z = \ln \pi + i(\frac{\pi}{2} + 2\pi k)$, $k \in \mathbb{Z}$
 (d) $z = \frac{\pi}{2} + 2\pi k \pm 4i$, $k \in \mathbb{Z}$
 (e) $z = \frac{\pi}{2} + \pi k$, $k \in \mathbb{Z}$
 (f) $z = \pi k i$, $k \in \mathbb{Z}$
 (g) $z = \pi k$, $k \in \mathbb{Z}$
 (h) $z = 2i$

3.50 $f'(z) = c z^{c-1}$

4.1 (a) 6
 (b) π

(c) 4
(d) $\sqrt{17}+\frac{1}{4}\sinh^{-1}(4)$

4.5 (a) $8\pi i$
(b) 0
(c) 0
(d) 0

4.6 (a) $\frac{1}{2}(1-i)$, $\frac{1}{2}(i-1)$, $-i$, 1
(b) πi, $-\pi$, 0, $2\pi i$
(c) $\pi i r^2$, $-\pi r^2$, 0, $2\pi i r^2$

4.7 (a) $\frac{1}{3}(e^3 - e^{3i})$
(b) 0
(c) $\frac{1}{3}(\exp(3+3i)-1)$

4.18 (a) $-4+i(4+\frac{\pi}{2})$
(b) $\ln(5) - \frac{1}{2}\ln(17) + i(\frac{\pi}{2} - Arg(4i+1))$
(c) $2\sqrt{2} - 1 + 2\sqrt{2}i$
(d) $\frac{1}{4}\sin(8) - 2 + i(2 - \frac{1}{4}\sinh(8))$

4.26 0 for $r < |a|$; $2\pi i$ for $r > |a|$

4.29 $\frac{2\pi}{\sqrt{3}}$

4.33 0

4.34 0 for $r=1$; $-\frac{\pi i}{3}$ for $r=3$; 0 for $r=5$

4.36 (a) $2\pi i$
(b) 0
(c) $-\frac{2\pi i}{3}$
(d) $\frac{2\pi i}{3}(e^3 - 1)$

5.1 (a) πi
(b) $-6\pi i$
(c) $4\pi i$
(d) 0

5.3 (a) 0
(b) $2\pi i$

(c) 0

(d) πi

(e) 0

(f) 0

7.1 (a) divergent

(b) convergent (limit 0)

(c) divergent

(d) convergent (limit $2 - \frac{i}{2}$)

(e) convergent (limit 0)

7.25 (a) $\sum_{k\geq 0}(-4)^k z^k$

(b) $\sum_{k\geq 0} \frac{1}{3\cdot 6^k} z^k$

(c) $\sum_{k\geq 0} \frac{k+1}{2\cdot 4^k} z^{k+2}$

7.26 (a) $\sum_{k\geq 0} \frac{(-1)^k}{(2k)!} z^{2k}$

(b) $\sum_{k\geq 0} \frac{(-1)^k}{(2k)!} z^{4k}$

(c) $\sum_{k\geq 0} \frac{(-1)^k}{(2k+1)!} z^{2k+3}$

(d) $\sum_{k\geq 1} \frac{(-1)^{k+1} 2^{2k-1}}{(2k)!} z^{2k}$

7.28 (a) $\sum_{k\geq 0}(-1)^k (z-1)^k$

(b) $\sum_{k\geq 1} \frac{(-1)^{k-1}}{k}(z-1)^k$

7.33 (a) ∞ if $|a| < 1$, 1 if $|a| = 1$, and 0 if $|a| > 1$

(b) 1

(c) 1

(d) 1

8.1 (a) $\{z \in \mathbb{C}: |z| < 1\}$, $\{z \in \mathbb{C}: |z| \leq r\}$ for any $r < 1$

(b) $\mathbb{C}$, $\{z \in \mathbb{C}: |z| \leq r\}$ for any r

(c) $\{z \in \mathbb{C}: |z-3| > 1\}$, $\{z \in \mathbb{C}: r \leq |z-3| \leq R\}$ for any $1 < r \leq R$

8.14 The maximum is 3 (attained at $z = \pm i$), and the minimum is 1 (attained at $z = \pm 1$).

8.17 One Laurent series is $\sum_{k\geq 0}(-2)^k (z-1)^{-k-2}$, converging for $|z-1| > 2$.

8.18 One Laurent series is $\sum_{k\geq 0}(-2)^k (z-2)^{-k-3}$, converging for $|z-2| > 2$.

8.19 One Laurent series is $-3(z+1)^{-1} + 1$, converging for $z \neq -1$.

8.25 (a) $\sum_{k\geq 0} \frac{(-1)^k}{(2k)!} z^{2k-2}$

8.36 (a) One Laurent series is $\sum_{k\geq -2} \frac{(-1)^k}{4^{k+3}}(z-2)^k$, converging for $0<|z-2|<4$.
(b) $-\frac{\pi i}{8}$

8.37 (a) $\sum_{k\geq 0} \frac{1}{e\,k!}(z+1)^k$
(b) $\frac{2\pi i}{e\,33!}$

9.5 (a) $2\pi i$
(b) $\frac{27\pi i}{4}$
(c) $-\frac{2\pi i}{17}$
(d) $\frac{\pi i}{3}$
(e) $2\pi i$
(f) 0

9.15 (c) $\frac{\pi}{2}$

9.21 (a) 0
(b) 1
(c) 4

Index

absolute convergence, 130
absolute value, 5
accumulation point, 13, 25
addition, 2
algebraically closed, 104
alternating harmonic series, 132
alternating zeta function, 190
analytic, 153
analytic continuation, 159
antiderivative, 78, 101, 196
Arg, 59
arg, 61
argument, 5
axis
    imaginary, 5
    real, 5

bijection, 32, 45
binary operation, 2
binomial coefficient, 190
boundary, 13, 118
boundary point, 13
bounded, 13
branch of the logarithm, 59

calculus, 1, 196
Casorati–Weierstraß theorem, 175
Cauchy's estimate, 152
Cauchy's integral formula, 87
    extensions of, 98, 152
Cauchy's theorem, 83
Cauchy–Goursat theorem, 83
Cauchy–Riemann equations, 34

chain of segments, 16
circle, 12
closed
    disk, 13
    path, 16
    set, 13
closure, 13
coffee, 89, 133, 175
comparison test, 128
complete, 125
complex number, 2
complex plane, 5
    extended, 48
complex projective line, 48
composition, 28
concatenation, 75
conformal, 32, 46, 119
conjugate, 10
connected, 14
continuous, 27
contractible, 85
convergence, 123
    pointwise, 132
    uniform, 132
convergent
    sequence, 123
    series, 127
cosine, 57
cotangent, 57, 194
cross ratio, 50
curve, 14
cycloid, 91

Dedekind sum, 194
dense, 175
derivative, 29
    partial, 33
difference quotient, 30
differentiable, 29
dilation, 46
discriminant, 18
disk
    closed, 13
    open, 12
    punctured, 170
    unit, 16
distance of complex numbers, 6
divergent, 123
domain, 24
double series, 158

e, 62, 125
embedding of $\mathbb{R}$ into $\mathbb{C}$, 3
empty set, 13
entire, 29, 105
essential singularity, 171
Euclidean plane, 12
Euler's formula, 8, 62
even, 169
exponential function, 56
exponential rules, 56
extended complex plane, 48

Fibonacci numbers, 191
field, 3
fixed point, 62
Frobenius problem, 192
function, 24
    conformal, 32, 46

    even, 169
    exponential, 56
    logarithmic, 59
    odd, 169
    trigonometric, 57
fundamental theorem
    of algebra, 4, 103, 166, 184, 188
    of calculus, 78, 101, 196

geogebra, 22
geometric interpretation of multiplication, 6
geometric series, 127
Green's theorem, 95, 197
group, 3

harmonic, 35, 111
harmonic conjugate, 113
holomorphic, 29
homotopic, 81
homotopy, 81
hyperbolic trig functions, 58

i, 4
identity map, 24
identity principle, 155
image
    of a function, 28
    of a point, 24
imaginary axis, 5
imaginary part, 4
improper integral, 105, 187
infinity, 47
inside, 89
integral, 72
    path independent, 103
integral test, 130

206 INDEX

integration by parts, 93
interior point, 13
inverse function, 32
    of a Möbius transformation, 45
inverse parametrization, 75
inversion, 46
isolated point, 13
isolated singularity, 171

Jacobian, 65
Jordan curve theorem, 89

L'Hôspital's rule, 198
Laplace equation, 111
Laurent series, 158
least upper bound, 125, 138
Leibniz's rule, 84, 197
length, 75
limit
    infinity, 47
    of a function, 25
    of a sequence, 123
    of a series, 127
linear fractional transformation, 44
Liouville's theorem, 105
Log, 60
log, 61
logarithm, 59
logarithmic derivative, 181

max/min property for harmonic functions, 116, 156
maximum
    strong relative, 116
    weak relative, 117, 156
maximum-modulus theorem, 156
mean-value theorem

for harmonic functions, 115
for holomorphic functions, 87
for real functions, 196
meromorphic, 182
minimum
    strong relative, 116
    weak relative, 156
minimum-modulus theorem, 156
Möbius transformation, 44
modulus, 5
monotone, 125
monotone sequence property, 125
Morera's theorem, 101
multiplication, 2

north pole, 52

obvious, 24
odd, 169
one-to-one, 32
onto, 32
open
    disk, 12
    set, 13
order of a pole, 174
orientation, 15

partial derivative, 33
path, 14
    closed, 16
    inside of, 89
    interior of, 89
    polygonal, 81
    positively oriented, 89
path independent, 103
periodic, 56, 194
Picard's theorem, 175

piecewise smooth, 73
plane, 12
pointswise convergence, 132
Poisson integral formula, 120
Poisson kernel, 96, 119
polar form, 9
pole, 171
polynomial, 4, 20, 41, 103
positive orientation, 89
power series, 136
    differentiation of, 148
    integration of, 140
primitive, 78
primitive root of unity, 9
principal argument, 59
principal logarithm, 59
principal value of a^b, 61
punctured disk, 170

real axis, 5
real number, 2
real part, 4
rectangular form, 9
region, 14
    of convergence, 137
    simply-connected, 102, 112
removable singularity, 171
reparametrization, 74
residue, 177
residue theorem, 177
reverse triangle inequality, 11, 20
Riemann hypothesis, 131
Riemann sphere, 48
Riemann zeta function, 131
root, 4
root of unity, 9

primitive, 9
root test, 139
Rouché's theorem, 183

separated, 14
sequence, 123
    convergent, 123
    divergent, 123
    limit, 123
    monotone, 125
series, 126
simple, 16
simply connected, 102
sine, 57
singularity, 170
smooth, 15
    piecewise, 73
south pole, 52
stereographic projection, 52

tangent, 57
Taylor series expansion, 149
topology, 12, 89
translation, 46
triangle inequality, 11
    reverse, 11
Tribonacci numbers, 192
trigonometric functions, 57
trigonometric identities, 7
trivial, 27

uniform convergence, 132
uniqueness theorem, 155
unit circle, 16
unit disk, 16
unit element, 3
unit sphere, 52

vector, 5

Weierstraß M-test, 135
Weierstraß convergence theorem, 164

CPSIA information can be obtained
at www.ICGtesting.com
Printed in the USA
BVHW04s1807080718
521075BV00003B/14/P